Contemporary Systems Thinking

Series Editor:
Robert L. Flood
Maastricht School of Management
The Netherlands

More information about this series at http://www.springer.com/series/5807

Gianfranco Minati • Mario R. Abram
Eliano Pessa
Editors

Towards a Post-Bertalanffy Systemics

Springer

Editors
Gianfranco Minati
AIRS/Italian Systems Society
Milano, Italy

Mario R. Abram
AIRS/Italian Systems Society
Milano, Italy

Eliano Pessa
University of Pavia
Pavia, Italy

ISSN 1568-2846
Contemporary Systems Thinking
ISBN 978-3-319-24389-4 ISBN 978-3-319-24391-7 (eBook)
DOI 10.1007/978-3-319-24391-7

Library of Congress Control Number: 2015959573

Springer Cham Heidelberg New York Dordrecht London

Printed on acid-free paper

Springer International Publishing AG Switzerland is part of Springer Science+Business Media (www.springer.com)

Preface

The title of the sixth national conference of the Italian Systems Society, "Towards a post-Bertalanffy Systemics", aims to underline the need for Systemics and Systems Science to generalize theoretically, interdisciplinarily and trans-disciplinarily using systemic concepts arising from the original or Bertalanffy Systemics, as well as from various disciplines themselves.

The topic of this sixth conference is an evolution of the subjects of previous conferences, namely:

- 2002 Emergence in Complex Cognitive, Social and Biological Systems
- 2004 Systemics of Emergence: Research and Applications
- 2007 Processes of Emergence of Systems and Systemic Properties—Towards a General Theory of Emergence
- 2011 Methods, Models, Simulations and Approaches Towards a General Theory of Change

Consideration was made of the generic first phase of Systemics devoted to overcoming classical mechanistic views, introducing new theoretical approaches studied, for instance, through Automata Theory, Catastrophe Theory, Chaos Theory, Control Theory, Cybernetics, Games Theory, Systems Dynamics, Gestalt, Sociobiology and Theory of Dynamical Systems. This phase can be characterized by the term "General System Theory", introduced by Ludwig von Bertalanffy (1901–1972) to generalize the concept of system by using some key systemic concepts such as interaction, general interdependence, openness and closeness, organization and homeostasis within the general framework of the isomorphism between sciences, searching for the unity of science. This phase continues but, over the past few years, two important cultural and scientific processes are occurring:

- After Systemics used the concept of system and related properties to overcome classical disciplines still tied to principles such as determinism, mechanistic view, summative assumption and linearity, reversibility, single optimum and equilibrium points, the disciplines themselves used in innovative ways the concept of system by introducing theoretical improvements. Examples are given by advances in disciplinary domains such as Theoretical Physics, Biology,

Neuroscience, Experimental Economics and Network Science, the latter being even a possible new version of Systemics itself due to its generality. The concepts and approaches considered by various disciplines using systemic approaches are extremely innovative and beg for their generalization.
- Phenomena are being considered and denoted in different ways, in various disciplines, using different approaches, but all are related to complexity, self-organization and emergence. However, all approaches considered by Systemics have a post-reductionist nature since they are unable to deal, for instance, with coherence and multiple coherences, dynamic structures, multiple models, non-homogeneity, nonequivalences, levels of distinguishability, multiple systems, power laws and scale-free properties.

Whence the original Systemics is suitable for dealing with processes of acquiring and maintaining the same or only a few, fixed systemic properties. Complex open systems, on the contrary, continuously acquire new, multiple, superimposed and often delocalized coherent sequences of properties.

Today it is extremely difficult to find disciplinary areas where the concept of system is not frequently used, albeit within specific contexts.

Indeed today disciplines have become sources, suppliers of new approaches, problems and systemic issues.

The interdisciplinary nature of the original Systemics and its power of generalization were given, overall, by the fact that the problems and solutions of one discipline become problems and solutions for another. Today, the modelling and interpretation of multidisciplinary approaches and representations facilitate this. The context, however, has changed dramatically.

This year's conference was devoted to identifying, discussing and understanding possible interrelationships of theoretical disciplinary improvements recognized as having prospective fundamental roles for a new post-Bertalanffy Systemics able to deal with problems of complexity in a generalized way where interdisciplinarity consists, for instance, in a disciplinary reformulation of problems, as from algebraic to geometrical, from military to political, from biological to chemical, and transdisciplinarity is related to the study of such reformulations and their properties. Examples of new issues introduced by such theoretical disciplinary improvements and studied within various disciplines include:

Between (the mesoscopic middle way)
Environment
Equivalence
Fractality
Individuality
Induction of properties
Irreversibility
Meta-structural properties
Methods and models to build strategies
Multiple, dynamic coherence
Mutation
Non-prescribability
Non-separability
Ontologies, scenarios and metamodels
Power laws
Pre-properties
Propagation
Quantum theories and concepts
Quasi properties
Quasiness
Symmetry
Structural dynamics

Networks
Non-causality
Nonequivalence
Non-invasiveness
Structural regimes of validity
System propagation
Topological dynamics
Transient

The new interdisciplinarity relates to properties of new representations, as in Network Science when dealing with topology, small worldness, power laws and fitness.

What is our role as a dynamical open network even though officially we are an association?

The challenge is still theoretical generalization and application, even where we have a lot of specificities, but know very little on how to combine them.

It is not simply replacing the old with the new, but to develop strategies to recognize, represent, model and act on new levels and combine, by considering, for example, multiple representations, functions and emergence.

In various disciplines this is already done, and inevitably well, since targets and projects are well specified and oriented.

The challenge is to do it for Systemics, with the vocations of cultural and theoretical generalization. The subject matter was explored through five sessions:

1. Studies of Emergence, Models and Simulations
2. The Contribution of Physics to a New General Theory of Systems
3. New Systemic Contents of Disciplinary Approaches and Problems
4. New Forms of Inter- and Trans-disciplinarity
5. Outlines of a New General Theory of Systems

The conference was opened with the plenary lecture by Emeritus Professor Fortunato Tito Arecchi, entitled "Quantum Effects in Linguistic Endeavors".

We conclude by mentioning that the Italian Systems Society does not want to perpetuate a role but, rather, contribute to the context-sensitive emergence of new, eventually collective, roles in an age where disciplines have a very high theoretical and applicative specialized systemic content and high interdisciplinarity but still require theoretical generalizations and suitable generalized approaches such as those studied by Network and Quantum Science as listed above. Systemics should also consider suitable cultural versions of such issues.

Milano, Italy — Gianfranco Minati
Milano, Italy — Mario R. Abram
Pavia, Italy — Eliano Pessa
April 2015

Acknowledgements

The sixth Italian conference has been possible thanks to the contributions of many people who have accompanied and supported the growth and development of AIRS during all the years since its establishment in 1985 and to the contribution of new energies from students and researchers realizing the systemic aspects of their activity.

We have been honoured by the presence of Professor Fortunato Tito Arecchi who delivered the opening plenary lecture.

We thank ARTEMIS NEUROSCIENCES, Rome, who sponsored the conference.

We thank the PONTIFICAL ATHENEUM S. ANSELMO for hosting the conference.

Thanks are also due to all the authors who submitted papers for this conference and in particular the members of the programme committee as well as the referees who have guaranteed the quality of the event.

We thank explicitly all the people who have contributed during the conference, bringing ideas and stimuli to this new phase of the scientific and cultural project of Systemics.

Contents

Part VI Outlines of a New General Theory of Systems

Editor Biography

Gianfranco Minati, Mathematician, Founder and president of the Italian Systems Society (AIRS); Doctoral lecturer at the Polytechnic of Milan; Member of the scientific committee of Conferences and Systems Societies. He is author of 32 chapters in books; editor of 7 books and journals; author or co-author of 15 books; author of 30 articles and of academic publications. His current research interest focuses on 1) Modelling processes of emergence by using Meta-Structures; 2) the emerging of a post-Bertalanffy Systemics; 3) the Dynamic Usage of Models (DYSAM), Logical Openness; 4) Architecture and Design as the design of social meta-structures to influence processes of emergence in social systems.

Mario R. Abram, Physicist, is an active member of the Italian Systems Society (AIRS). He worked mainly in ENEL (Italian Power Agency), at Research Department (Automatica Research Center), then in Cesi S.p.A. and Cesi Ricerca S.p.A. He experienced hybrid and digital simulation systems, working on models of power systems, thermoelectric and nuclear power plants and processes control. He participated in the building of real-time simulators for thermoelectric power plants and their application to control systems testing and tuning, in the development of supervision and control systems for power testing facilities and distribution networks. Also he worked on the application of neural networks to power systems supervision and investigated safety conditions in interactions between infrastructures. He authored several publications in systems science. His research interests include: dynamical systems, modelling and simulation of processes and interactions between infrastructural networks.

Eliano Pessa, Theoretical Physicist, is actually Full Professor of General Psychology and Cognitive Modeling at the University of Pavia, Italy. He has already been Dean of the Department of Psychology and of the Inter-departmental Research Center on Cognitive Science in the same university. In the past he has been Associate Professor of Artificial Intelligence at the University of Rome "La Sapienza", Faculty of Psychology. He is author or co-author of 10 books and of a large number of papers on scientific journals, books, proceedings of international conferences. His scientific research interests include: quantum theories of brain operation, computational neuroscience, artificial neural networks, models of emergence processes, quantum field theory, models of phase transitions in condensed matter, models of human memory and visual perception, models of decision making, models of statistical reasoning.

Scientific Committee

G. Minati (chairman)	Italian Systems Society
E. Pessa (co-chairman)	University of Pavia
S. Di Gregorio	University of Calabria
C. Fontana	Polytechnic University of Milan
G. Bruno	ISIA Roma Design
G. Vitiello	University of Salerno
I. Licata	School of Advanced International Studies on Theoretical and Non Linear Methodologies of Physics, Bari
L. Biggiero	University of L'Aquila
M.P. Penna	University of Cagliari
R. Serra	University of Modena and Reggio Emilia

List of Contributors

Mario R. Abram
AIRS (Associazione Italiana per la Ricerca sui Sistemi), Milano, Italy

Alessandra Addis
Department of Pedagogy, Psychology, Philosophy, Faculty of Humanistic Studies, University of Cagliari, Cagliari, Italy

Mirian Agus
Department of Pedagogy, Psychology, Philosophy, Faculty of Humanistic Studies, University of Cagliari, Cagliari, Italy

Fortunato T. Arecchi
University of Firenze and INO-CNR, Largo Enrico Fermi, Firenze, Italy

Pier Luca Bandinelli
Dipartimento di Scienze Umane (Comunicazione, Formazione, Psicologia), Libera Università LUMSA, Roma, Italy

Dipartimento di Salute Mentale RMA, Roma, Italy

AIRS - Associazione Italiana per la Ricerca sui Sistemi, Milano, Italy

Marta Bertolaso
Faculty of Engineering, Institute of Philosophy of Scientific and Technological Practice, University Campus Bio-Medico di Roma, Roma, Italy

Leonardo Bich
Department of Logic and Philosophy of Science, IAS-Research Centre for Life, Mind and Society, University of the Basque Country (UPV/EHU), Donostia-San Sebastiàn, Spain

Giordano Bruno
ISIA Roma Design, Istituto Superiore per le Industrie Artistiche, Roma, Italy

Alessandra Cucurnia
Dipartimento di Architettura, Università di Firenze, Florence, Italy

Valerio De Cecio
ISIA Roma Design, Piazza della Maddalena 53, 00196 Rome, Italy
Link Campus University, Via Nomentana 335, 00162 Rome, Italy

Piero De Giacomo
Department of Neurological and Psychiatric Sciences, University of Bari, Bari, Italy

Valerio Di Battista
Politecnico di Milano, Milano, Italy
Rete Piemontese degli Osservatori del Paesaggio, Piedmont Network of Landscape Observatories, Provincia di Alessandria, Italy

Mirko Di Bernardo
Dipartimento di Studi di Impresa Governo Filosofia, Università degli Studi di Roma "Tor Vergata", Roma, Italy

Umberto Di Caprio
ASDE (Associazione Dirigenti Enel), Roma, Italy

Nicola Di Stefano
Institute of Philosophy of Scientific and Technological Practice, University Campus Bio-Medico di Roma, Roma, Italy

Maria Chiara Fastame
Department of Pedagogy, Psychology, Philosophy - Faculty of Humanistic Studies, University of Cagliari, Via Is Mirrionis 1, 09123 Cagliari, Italy

Alessandro Filisetti
Current address: Explora s.r.l, Rome, Italy
European Centre for Living Technology, Ca Minich, S. Marco, Venezia, Italy

Rodolfo A. Fiorini
Department of Electronics, Information and Bioengineering (DEIB), Politecnico di Milano University, Piazza Leonardo da Vinci, Milano, Italy

Carlotta Fontana
Dipartimento di Architettura e Studi Urbani (DAStU), Politecnico di Milano, Milano, Italy

Giorgio Fumera
Department of Electrical and Electronic Engineering, Piazza d'Armi S.n.c., Cagliari, Italy

Giorgio Giallocosta
Dipartimento di Scienze per l'Architettura, Università di Genova, Stradone S. Agostino, Genoa, Italy

Marco Giunti
ALOPHIS, Applied Logic Language Philosophy and History of Science, University of Cagliari, Cagliari, Italy

Roberta Grimaldi
ISIA Roma Design, Piazza della Maddalena 53, 00196 Rome, Italy
Link Campus University, Via Nomentana 335, 00162 Rome, Italy

Paul Kenneth Hitchcott
Department of Psychology, Southampton Solent University, Southampton, Hampshire, UK

Pier Luigi Marconi
Dipartimento di Psicologia Dinamica e Clinica, Università di Roma "La Sapienza", Roma, Italy
ARTEMIS Neurosciences StP, Roma, Italy
AIRS - Associazione Italiana per la Ricerca sui Sistemi, Milano, Italy

Maria Lidia Mascia
Department of Pedagogy, Psychology, Philosophy, Faculty of Humanistic Studies, University of Cagliari, Cagliari, Italy

Mia A. Massarini
Istituto Comprensivo di Castel San Giovanni e Sarmato "Cardinale Agostino Casaroli", Castel San Giovanni (PC), Italy

Claudio Mazzola
School of History, Philosophy, Religion and Classics, The University of Queensland, Forgan Smith Building (1), Brisbane, QLD, Australia

Carlo Maria Medaglia
DIAG, Sapienza Università di Roma, Piazzale Aldo Moro, Rome, Italy

Gianfranco Minati
Italian Systems Society, Milan, Italy

Eraldo Nicotra
Dipartimento di Pedagogia, Psicologia, Filosofia, University of Cagliari, Cagliari, Italy

Antonio Opromolla
ISIA Roma Design, Piazza della Maddalena 53, 00196 Rome, Italy
Link Campus University, Via Nomentana 335, 00162 Rome, Italy

Mauro Palatucci
ISIA Roma Design, Piazza della Maddalena, Rome, Italy

Maria Pietronilla Penna
Dipartimento di Pedagogia, Psicologia, Filosofia - Università di Cagliari, Cagliari, Italy

Silvia Perini
Dipartimento di Lettere, Arti, Storia e Società, Università degli Studi di Parma, Parma, Italy

Eliano Pessa
Dipartimento di Scienze del Sistema Nervoso e del Comportamento, Università di Pavia, Pavia, Italy

AIRS - Associazione Italiana per la Ricerca sui Sistemi, Milano, Italy

Emanuela Pietrocini
Early Music Department Schola Palatina, La Vertuosa Compagnia de' Musici di Roma, ISIA of Pescara, Multimedia Design, Pescara, Italy

Simone Pinna
Alophis – Applied Logic Language Philosophy and History of Science, University of Cagliari, Via is Mirrionis 1, 09123 Cagliari, Italy

Andrea Roli
Department of Engineering and Computer Science (DISI), Alma Mater Studiorum University of Bologna, Via Venezia 52, I-47521 Cesena, Italy

European Centre for Living Technology, Ca Minich, S. Marco 2940, 30124 Venezia, Italy

Dolores Rollo
Dipartimento di Neuroscienze, Università degli Studi di Parma, Parma PR, Italy

Giulia Romiti
ISIA Roma Design, Istituto Superiore per le Industrie Artistiche, Roma, Italy

Giulia F. Santacroce
BV-Tech (S.p.A), Milano, Italy

Roberto Serra
Department of Physics, Informatics and Mathematics, University of Modena e Reggio Emilia,
Via Campi 213b, 41125 Modena, Italy

European Centre for Living Technology, Ca Minich, S. Marco 2940, 30124 Venezia, Italy

Andrea Spoto
Department of General Psychology, University of Padua, Padua, Italy

Francesco Sulla
Dipartimento di Lettere, Arti, Storia e Società, Università degli Studi di Parma, Parma PR, Italy

Guido Tascini
Centro Studi G.B. Carducci, Corso Marconi, Fermo, Italy

Marco Villani
Department of Physics, Informatics and Mathematics, University of Modena e Reggio Emilia, Via Campi 213b, 41125 Modena, Italy

European Centre for Living Technology, Ca Minich, S. Marco 2940, 30124 Venezia, Italy

Giuseppe Vitiello
Dipartimento di Fisica "E.R. Caianiello", Università di Salerno and INFN, Sezione di Napoli, Gruppo Collegato Salerno, Fisciano (Salerno), Italy

Valentina Volpi
ISIA Roma Design, Piazza della Maddalena 53, 00196 Rome, Italy

Link Campus University, Via Nomentana 335, 00162 Rome, Italy

Part I
Opening Lecture

Chapter 1
Quantum Effects in Linguistic Endeavors

Fortunato T. Arecchi

1 A Methodological Premise

Any scientific description entails a complementarity between its Extension and Detail. The intrinsic limitation of a TOE (Theory Of Everything) is its inability to predict the detailed operations of single individuals, even though it provides global explanations. The two aspects of a scientific description, namely, Extension (E) and Detail (D), result mutually conflicting; this may be symbolically condensed by an uncertainty relation as

$$\Delta E \times \Delta D > C, \tag{1.1}$$

where Δ denotes the amount of uncertainty and C is a suitable constant related to the descriptive theory. For instance, a successful approach explaining the cosmic evolution from the Microwave Background to the Galaxy formation can not explain the details of planet differentiation, why e.g. the Earth has a magnetic field providing the Van Allen belt shield from solar particles, or the water necessary for life. Two different sets of foundational principles must be introduced in order to explain the two classes of phenomena, that is, the fundamental objects of Planetology and their mutual interactions must be introduced appropriately and cannot be derived from the general principles of Cosmology.

In a similar way, a general powerful QFT (Quantum Field Theory) approach has been developed to explain the brain and memory organization starting from the collective organization of water dipole quanta in living matter [1, 2].

Even though, the puzzling fact that in a human linguistic endeavor, words are mutually influencing through their meanings, so that an "infinite use emerges from

F.T. Arecchi (✉)
University of Firenze and INO-CNR, Largo Enrico Fermi 6, 50125 Firenze, Italy
e-mail: tito.arecchi@ino.it

G. Minati et al. (eds.), *Towards a Post-Bertalanffy Systemics*, Contemporary Systems Thinking, DOI 10.1007/978-3-319-24391-7_1

a finite amount of resources" [3] has no explanation whatsoever in the QFT of brain and memory, that has no tools to differentiate the human brain behavior from that of other animals.

In this work, we present a specific theoretical approach that provides that sound explanation for the linguistic performances not achievable in the QFT of brain and memory.

2 Two Separate Cognitive Processes

In [4] I have analyzed two distinct moments of human cognition, namely, *apprehension* (A) whereby a coherent perception emerges from the recruitment of neuronal groups, and *judgment* (B) whereby memory recalls previous (A) units coded in a suitable language, these units are compared and from comparison it follows the formulation of a judgment.

The first moment (A) has a duration around 1 s; its associated neuronal correlate consists of the synchronization of the EEG (Electro-Encephalo-Graphic) signals in the so-called gamma band (frequencies between 40 and 60 Hz) coming from distant cortical areas. It can be described as an interpretation of the sensorial stimuli on the basis of available algorithms, through a Bayes inference.

Precisely [4], calling h ($h =$ hypothesis) the interpretative hypotheses in presence of a sensorial stimulus d ($d =$ datum), the Bayes inference selects the most plausible hypothesis h^*, that determines the motor reaction, exploiting a memorized algorithm $P(d|h)$, that represents the conditional probability that a datum d be the consequence of an hypothesis h. The $P(d|h)$ have been learned during our past; they represent the equipment whereby a cognitive agent faces the world. By equipping a robot with a convenient set of $P(d|h)$, we expect a sensible behavior.

The second moment (B) entails a comparison between two apprehensions (A) acquired at different times, coded in a given language and recalled by the memory. If, in analogy with (A), we call d the code of the second apprehension and h^* the code of the first one, now—at variance with (A)—h^* is already given; instead, the relation $P(d|h)$ which connects them must be retrieved, it represents the conformity between d and h^*, that is, the best interpretation of d in the light of h^*. Thus, in linguistic operations, we compare two successive pieces of the text and extract the conformity of the second one on the basis of the first one. This is very different from (A), where there is no problem of conformity but of plausibility of h^* in view of a motor reaction. Let us make two examples: a rabbit perceives a rustle behind a hedge and it runs away, without investigating whether it was a fox or just a blow of wind. On the contrary, to catch the meaning of the 4-th verse of a poem, we must recover the meaning of the 3-d verse of that same poem, since we do not have a-priori algorithms to provide a satisfactory answer. Once the judgment, that is, the $P(d|h)$ binding the codes of the two linguistic pieces in the best way, has been built, it becomes a memorized resource to which to recur whenever that text is presented again. It has acquired the status of the pre-learned algorithms that rule (A). However—at variance with mechanized resources—whenever we re-read the same poem, we can grasp new meanings that enrich the previous judgment $P(d|h)$. As in any exposure to a text (literary, musical, figurative) a re-reading improves our understanding. (B) requires about 3 s and entails *self-consciousness*, as the agent who expresses the judgment must be aware that the two successive apprehensions are both under his/her scrutiny and it is up to him/her to extract the mutual relation. As a fact, exploring human subjects with sequences of simple words, we find evidence of a limited time window around 3 s [5, 6], corresponding to the memory retrieval of a linguistic item in order to match it with the next one in a text flow (be it literary, or musical, or figurative).

At variance with (A), (B) does not presuppose an algorithm, but rather it builds a new one through an ***inverse Bayes procedure*** [7]. This construction of a new algorithm is a sign of *creativity and decisional freedom*. Here the question emerges: can we provide a computing machine with the (B) capacity, so that it can emulate a human cognitive agent? [8]. The answer is NOT, because (B) entails non-algorithmic jumps, insofar as the inverse Bayes procedure generates an ad hoc algorithm, by no means pre-existent.

The five figures that follow and their captions explore in detail these aspects [4] (Figs. 1.1, 1.2, 1.3, 1.4 and 1.5).

After having shown evidence of this short term memory window bridging successive pieces of a linguistic text, we formulate a quantum conjecture. This conjecture fulfills two needs, namely, (1) explaining the fast search in a semantic space, whose sequential exploration by classical mechanisms would require extremely long times, incompatible with the cadence of a linguistic presentation [9]; (2) introducing a fundamental uncertainty ruled by a quantum constant that yields a decoherence time fitting the short term memory window. The memory enhancement associated with linguistic flows is an exclusively human operation, not applicable to a cognitive agent that operates ***recursively***, exploiting algorithms already stored in the memory. If the conjecture will be confirmed, the quantum mechanism would explain the a-posteriori construction of novel interpretational tools.

Classifying the information content of spike trains, an uncertainty relation emerges between the bit size of a word and its duration. This uncertainty is ruled by a quantum constant that can be given a numerical value and that has nothing to do with Planck's constant. A quantum conjecture might explain the onset and decay of the memory window connecting successive pieces of a linguistic text. The conjecture here formulated is applicable to other reported evidences of quantum effects in human cognitive processes, so far lacking a plausible framework since no efforts to assign a quantum constant have been associated.

Models of quantum behavior in language and decision taking have already been considered by several Authors but without a dynamical basis, starting from 1995 [10, 11]; and over the past decade [12]. Most references are collected in a recent book [13]. None of these Authors worries about the quantum constant that must replace Planck's constant. However, a quantum behavior entails pairs of incompatible variables, whose measurement uncertainties are bound by a quantization constant, as Planck's in the original formulation of Heisenberg. One can not apply a quantum formalism without having specified the quantum constant rul-

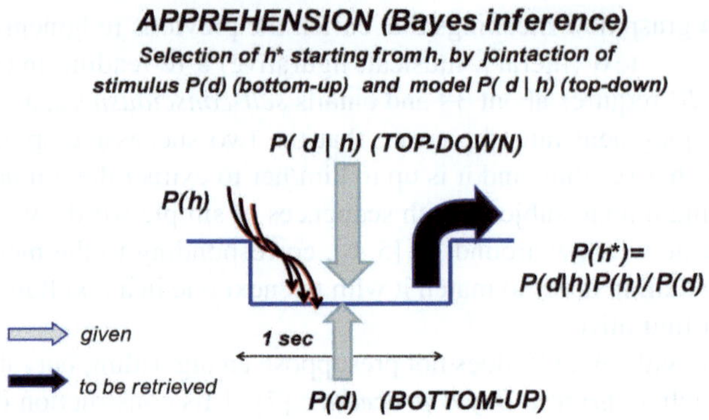

Fig. 1.1: Apprehension as a Bayes inference. One formulates a manifold of hypotheses; each one provides a datum through the top-down conditional probability; only the hypothesis that generates the actual datum (bottom-up) is the plausible one

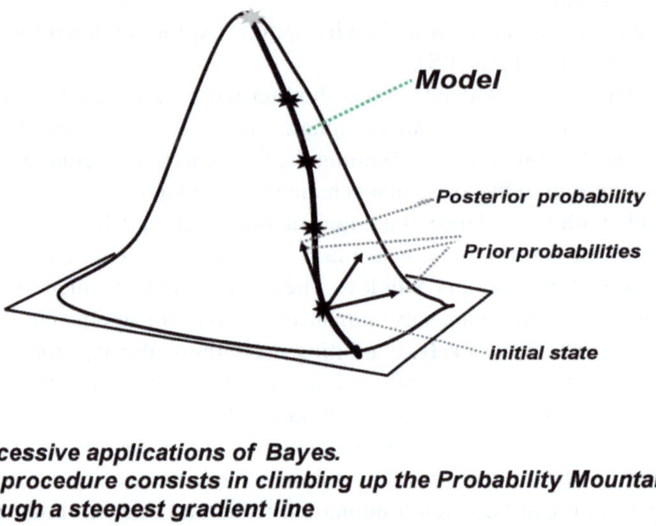

Fig. 1.2: Recursive application of Bayes is equivalent to climbing a probability mountain, guided by the Model, that is, the conditional probability that an hypothesis generates a datum. This strategy is common e.g. to Darwin evolution and to Sherlock Holmes criminal investigation; since the algorithm is unique, it can be automatized in a computer program (expert system)

ing the formalism. For this reason, all reported quantum attempts must be considered flawed, because (1) either they overlook the need for a quantization constant [10–12], or (2) use Planck constant and consequently arrive to very short decoherence times, incompatible with cognitive processes [13–16].

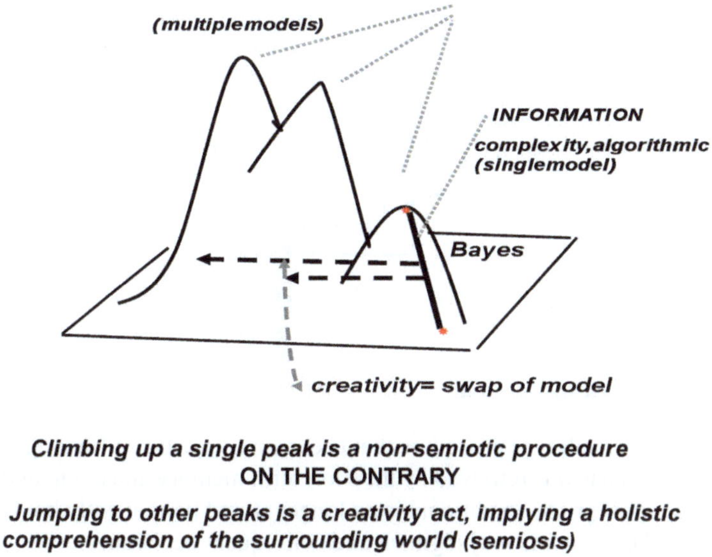

Fig. 1.3: Comparison of two different complexities, namely, (1) the algorithmic complexity, corresponding to the bit length of the program that enables the expert system to a recursive Bayes; and (2) semantic complexity, corresponding to the occurrence of different models (provided they are countable); in fact they are not, because we will see that different meanings result from a quantum exploration

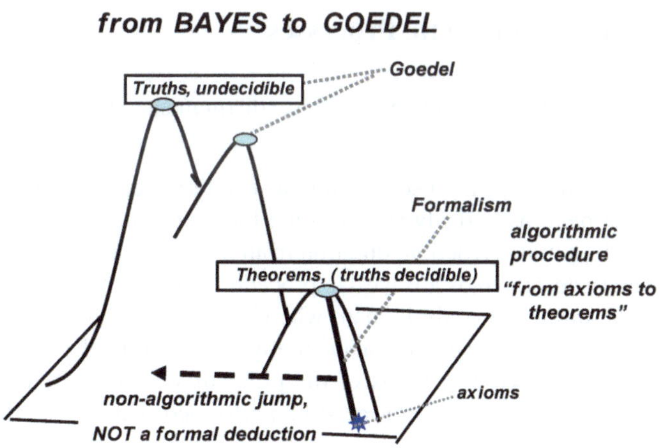

Fig. 1.4: The semantic complexity explains the criticism of K. Gödel to Hilbert's formal deduction of theorems as the only way to extract truths from a body of axioms

After summarizing in Sect. 3 the main difference between (A)-perception and (B)-linguistic processes, we devote Sect. 4 to the quantum aspects of an interrupted

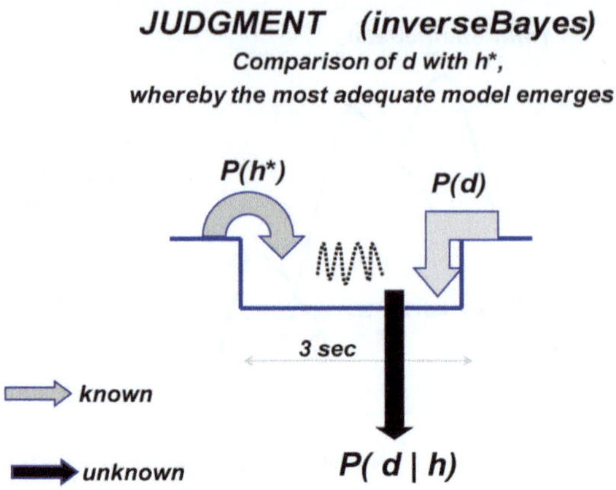

Fig. 1.5: The inverse Bayes procedure that occurs in linguistic endeavors, whereby a previous piece of a text is retrieved by the short term memory and compared with the next one: the appropriate conditional is no longer stored permanently but it emerges as a result of the comparison (judgment and consequent decision)

spike train, that provide a non-Newtonian quantization suitable for the foundation of quantum linguistic processes.

3 Perceptions vs Linguistic Processes

To summarize the previous arguments, we have distinguished between two different cognitive processes, namely,

- (*A*) *perception*, whereby a sensorial stimulus is interpreted in terms of "models", or behavioral patterns, already stored in the long term memory; the interpreted stimulus elicits a motor reaction; duration from a few hundred milliseconds up to 1 s; adequately described as a Bayesian procedure; common to all animals, and
- (*B*) *linguistic processes*, only human, whereby a sequence of pieces, coded as words of the same language, are sequentially presented to the cognitive agent; each piece is interpreted in terms of the previous one recovered by the short term memory; such a comparison must be performed within 3 s; otherwise, the sequence must be repeated.

Focusing on (*B*), a decision, or *judgment*, is the interpretation of the last piece based upon the meanings of the previous one. Scanning all possible meanings of each piece entails a fast search process that requires a quantum search.

Plenty of approaches have tackled quantum-like aspects of language processing [10–13]; however these approaches either did not discuss limitations due to a quantum constant, hence, they are purely formal without a physical basis or they refer to the quantum behavior of Newtonian particles [14, 15] and hence are limited by a coherence time estimated around 10^{-14} s, [16, 17] well below the infra-sec scale of the cognitive processes [18].

4 A Novel Aspect of Quantum Behavior

Standard quantum physics emerges from the Newtonian physics of a single particle. Refer for simplicity to 1-dimension. The uncertainties of position x and momentum p obey the Heisenberg condition

$$\Delta x \Delta p \geq \hbar. \tag{1.2}$$

All quantum formalism is a consequence. For instance, comparison with the Fourier condition

$$\Delta x \Delta k \geq 1. \tag{1.3}$$

suggests the De Broglie relation

$$k = \frac{p}{\hbar}, \tag{1.4}$$

whence the single particle interference, which contains the only quantum mystery [19] and Schrödinger wave equation. In the x, p space, instead of Euclidean points $\Delta x = 0$, $\Delta p = 0$, we have uncertainty rectangles; thus the uncertainty areas of two separate particles can overlap: this is the origin of *entanglement*.

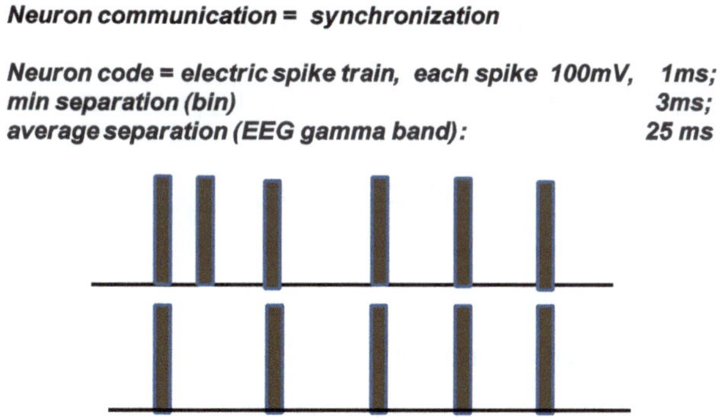

Fig. 1.6: Two neural spike trains; synchronization missed for an extra-spike in the upper train

Now, let us consider a non-Newtonian phenomenon consisting of a temporal train of identical spikes of unit area and duration $\tau_b = 3\,\text{ms}$ (bin) positioned at unequal times. This is a sound model for the electrical activity of a cortical neuron [20, 21]. The corresponding signal is a binary sequence of 0's and 1's, depending on whether a given bin is empty or filled by a spike. Spike synchronization, i.e. temporal coincidence of 0's and 1's, is considered as the way cortical neurons organize in a collective state [22] (Fig. 1.6).

Each cortical neuron has two ways to modify its spike occurrence, namely, either coupling to other cortical neurons or receiving signals from extra-cortical regions.

Let us take a processing time $T = 300\,\text{ms}$, then, the total number of binary words that can be processed is $P_M = 2^{300/3} \cong 10^{33}$. At the end of a computational task a decision center (called GWS = global workspace [23, 24]) picks up the information of the largest synchronized group and—based upon it—elicits a decision.

In the ***perceptual*** case (A), the cognitive action combines a bottom-up signal provided by the sensorial organs with a top-down interpretation provided by long term memories stored in extra-cortical areas [25].

In the ***linguistic*** case (B), the comparison occurs between the code of the second piece and the code of the previous one retrieved by the short term memory. Here, we should consider a fact which so far had escaped a full explanation. Namely,spikes occur at average rates corresponding to the so called EEG gamma band (say, around 50 Hz, that is, average separation 20 ms) [20]. However, superposed to the gamma band, there is a low frequency background (theta band, around 7 Hz), which controls the number of gamma band bursts [26]. We show that interruption of a spike train introduces a quantum uncertainty, hence an entanglement among different words. This entanglement provides a fast quantum search of meanings, that in classical terms would take a much longer time.

The theta-gamma cross-modulation corresponds to stopping the neural sequence at $\Delta T \leq T$. As a result, all spike trains equal up to ΔT, but different by at least one spike in the interval $T - \Delta T$, provide an uncertainty cloud ΔP such that [27, 28]

$$\Delta P = 2^{(T-\Delta T)/\tau} = P_M\, 2^{-\Delta T/\tau} \tag{1.5}$$

Thus we have a peculiar uncertainty of exponential type between spike information P and duration T, that is,

$$\Delta P \cdot 2^{\Delta T/\tau} = P_M \tag{1.6}$$

By a variable change

$$y = 2^T \tag{1.7}$$

we arrive to a product type uncertainty relation

$$\Delta P \Delta y = P_M \tag{1.8}$$

In the space (P, y) we have a Heisenberg-like uncertainty relation. Following the standard procedure of a quantum approach, we expect single particle interference and two particle entanglement in such a space.

For $\Delta P = 1$ (minimal disturbance represented by 1 spike) we have the *decoherence* pseudo time $\Delta y_d = P_M$. The corresponding decoherence time (in bins) is

$$decoherence\ time = \log_2 P_M = 100\ (bins) \tag{1.9}$$

and going from bins to sec:

$$decoherence\ time = 0.3\,\text{s} \tag{1.10}$$

very far from the naive value of 10^{-14} s evaluated for Newtonian particles disturbed by the thermal energy $k_B T$ at $T =$ room temperature [16].

5 Conclusions

To conclude, we stress the revolution brought about by the linguistic processes in the brain:

1. The quantum constant for spike number-duration uncertainty has nothing to do with Planck's constant, a new type of quantum behavior has to be considered; spike synchronization is a peculiar physical process that cannot be grasped in terms of Newtonian position-momentum.
2. The energy disturbance which rules the decoherence time is by no means $k_B T$ (k_B being Boltzmann constant and T the room temperature), but it is replaced by the minimal energy necessary to add or destroy a cortical spike. This energy corresponds to the opening along the axon of about 10^7 ionic channels each one requiring an $ATP \to ADP + P$ reaction involving 0.3 eV, thus the minimal energy disturbance in neural spike dynamics is around $10^8 k_B T$ [29]. This is the

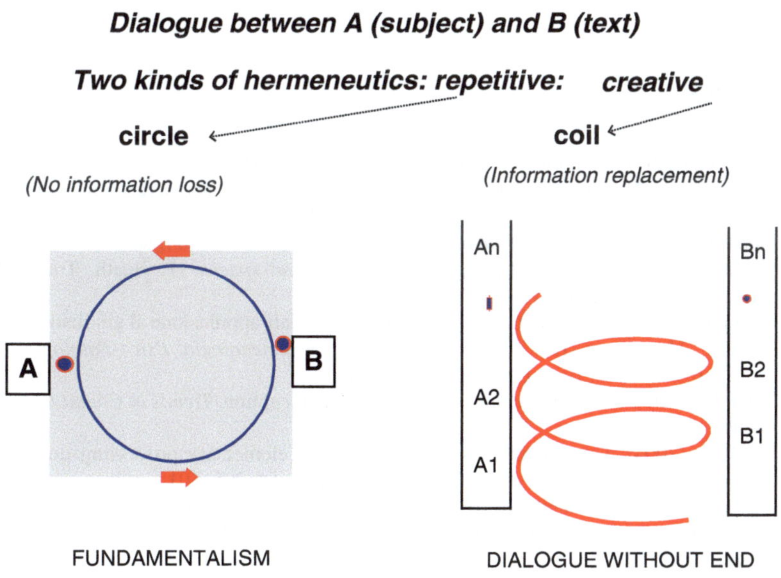

Fig. 1.7: Visual comparison between two kinds of interpretation of a text, or hermeneutics, namely, the CIRCLE, whereby the interpreter A attributes a finite and fixed set of meanings to the text B, and the COIL, whereby A captures some particular aspects of B and—based on that information—A approaches again the text B discovering new meanings

evolutionary advantage of a brain: to live comfortably at room temperature and be barely disturbed, as it were cooled at 10^8 the room temperature.

As for the interpretation (hermeneutics) of a cognitive experience (be it perceptual or linguistic), we represent in Fig. 1.7 the procedural interpretation by a computing machine (CIRCLE) against that of any human language (COIL).

As for the CIRCLE, in information science, an ***ontology*** is a formal definition of the properties, and interrelationships of the entities that exist for a particular domain of discourse. An ontology compartmentalizes the variables needed for some set of computations and establishes the relationships between them. For instance, the booklet of the replacement parts of a brand of car is the ontology of that car. The fields of artificial intelligence create ontologies to limit complexity and to organize information. The ontology can then be applied to problem solving. Nothing is left out; we call this cognitive approach "finitistic" as nothing is left out beyond the description.

On the contrary, in any human linguistic endeavor (be it literary, or musical or figurative) A starts building a provisional interpretation $A1$ of the text; whenever A returns to B, he/she has already some interpretational elements to start with, and from there A progresses beyond, grasping new aspects $B2$, $B3$ and hence going to $A2$ and so on (COIL). If B is not just a linguistic text, but another human subject, then B undergoes similar hermeneutic updates as A; this is a picture of the dialogical exchange between two human beings.

References

1. Vitiello, G. (2001). *My double unveiled*. Amsterdam: John Benjamins.
2. Freeman, W., & Vitiello, G. (2006). Nonlinear brain dynamics as macroscopic manifestation of underlying many-body dynamics. *Physics of Life Reviews, 3*, 93–118.
3. Humboldt, W. (1836). *On language: On the diversity of human language construction and its influence on the mental development of the human species* (P. Heath, Trans., 1988). Cambridge: Cambridge University Press.
4. Arecchi, F. T. (2012). Fenomenologia della coscienza: dall'apprensione al giudizio. In A. Ales Bello & P. Manganaro (Eds.), *... e la coscienza? Fenomenologia, Psico-Patologia, Neuroscienze* (pp. 841–875). Bari: Ed. Giuseppe Laterza.
5. Pöppel, E. (1997). A hierarchical model of temporal perception. *Trends in Cognitive Sciences, 1*, 56–61.
6. Pöppel, E. (2004). Lost in time: A historical frame, elementary processing units and the 3-second window. *Acta Neurobiologiae Experimentalis, 64*, 295–301.
7. Arecchi, F. T. (2007). Complexity, Information Loss and Model Building: From neuro- to cognitive dynamics. In *SPIE Noise and Fluctuation in Biological, Biophysical, and Biomedical Systems*. Paper 6602-36.
8. Turing, A. (1950). Computing machinery and intelligence. *Mind, 59*, 433–460.
9. Grover, L. K. (2001). From Schrödinger's equation to quantum search algorithm. *American Journal of Physics, 69*, 769–777.
10. Aerts, D., & Aerts, S. (1995). Applications of quantum statistics in psychological studies of decision processes. *Foundations of Science, 1*, 85–97.
11. Aerts, D. (2009). Quantum structure in cognition. *Journal of Mathematical Psychology, 53*(5), 314–348.

12. Khrennikov, A. Y. (2010). *Ubiquitous quantum structure: From psychology to finance*. New York: Springer.
13. Busemeyer, J. R., & Bruza, P. D. (2012). *Quantum models of cognition and decision*. Cambridge: Cambridge University Press.
14. Penrose, R. (1994). *Shadows of the mind*. New York: Oxford University Press.
15. Hagan, S., Hameroff, S. R., & Tuszynski, J. A. (2002). Quantum computation in brain microtubules: Decoherence and biological feasibility. *Physical Review E, 65*, 061901.
16. Tegmark, M. (2000). The importance of quantum decoherence in brain processes. *Physical Review E, 61*, 4194–4206.
17. Koch, C., & Hepp, K. (2006). Quantum mechanics in the brain. *Nature, 440*, 611.
18. Rodriguez, E., George, N., Lachaux, J.P., Martinerie, J., Renault, B., & Varela, F. (1999). Perception's shadow: Long-distance synchronization in the human brain. *Nature, 397*, 340–343.
19. Feynman, R. (1966). *The Feynman lectures on physics* (Vol. 3, Chapter 1).
20. Rieke, F., Warland, D., de Ruyter van Steveninck, R., & Bialek, W. (1996). *Spikes: Exploring the neural code*. Cambridge: MIT Press
21. Arecchi, F. T. (2004). Chaotic neuron dynamics, synchronization and feature binding. *Physica A, 338*, 218–237.
22. Singer, W., & Gray, C. M. (1995). Visual feature integration and the temporal correlation hypothesis. *Annual Reviews of Neuroscience, 18*, 555–586.
23. Baars, B. J. (1988). *A cognitive theory of consciousness*. Cambridge: Cambridge University Press.
24. Dehaene, S., Sergent, C., & Changeux, J.-P. (2003). A neuronal network model linking subjective reports and objective physiological data during conscious perception. *Proceedings of the National Academy of Sciences of the United States of America, 100*(14), 8520–8525.
25. Grossberg, S. (1995). The attentive brain. *American Scientist, 83*, 438–449.
26. Jensen, O., & Colgin, L. L. (2007). Cross-frequency coupling between neuronal oscillations. *Trends in Cognitive Science, 11*, 267–269.
27. Arecchi, F. T. (2003). Chaotic neuron dynamics, synchronization and feature binding: Quantum aspects. *Mind and Matter, 1*, 15–43.
28. Arecchi, F. T. (2004). *Caos e complessità nel vivente* (p. 248). Pavia: IUSS Press.
29. Arecchi, F. T. (2013). La coscienza e il linguaggio: perché e come. Congetture quantistiche. In L. Urbani Ulivi (Ed.), *Strutture di Mondo. Il pensiero Sistemico come Specchio di una Realtà Complessa* (Vol. 2, pp.199–226). Bologna: Il Mulino.

Part II
Studies of the Emergence: Models and Simulations

Chapter 2
Cross-Frequency Modulation, Network Information Integration and Cognitive Performance in Complex Systems

Pier Luigi Marconi, Pier Luca Bandinelli, Maria Pietronilla Penna, and Eliano Pessa

1 The Systemic Approach to the Study of Brain Activity

A consistent body of neurophysiological research, performed in the last 20 years (and particularly in the last decade) has shown that the higher cognitive processes can not be explained by processes of sequential neural processing. As well known, past neurophysiologic studies based on the technique of evoked potentials were based on this assumption. However, the actually available faster, portable and powerful computational tools allowed the development of far more sophisticated signal processing algorithms. This circumstance led to the availability of the technical and methodological background needed for the study of electroencephalographic

P.L. Marconi
Dipartimento di Psicologia Dinamica e Clinica, Università di Roma "La Sapienza", Roma, Italy

ARTEMIS Neurosciences StP, Roma, Italy

AIRS - Associazione Italiana per la Ricerca sui Sistemi, Milano, Italy
e-mail: pierluigi.marconi@uniroma1.it

P.L. Bandinelli
Dipartimento di Scienze Umane (Comunicazione, Formazione, Psicologia), Libera Università LUMSA, Roma, Italy

Dipartimento di Salute Mentale RMA, Roma, Italy

AIRS - Associazione Italiana per la Ricerca sui Sistemi, Milano, Italy

M.P. Penna
Dipartimento di Pedagogia, Psicologia, Filosofia, Università di Cagliari, Cagliari, Italy

AIRS - Associazione Italiana per la Ricerca sui Sistemi, Milano, Italy

E. Pessa (✉)
Dipartimento di Scienze del Sistema Nervoso e del Comportamento,
Università di Pavia, Pavia, Italy

AIRS - Associazione Italiana per la Ricerca sui Sistemi, Milano, Italy
e-mail: eliano.pessa@unipv.it

G. Minati et al. (eds.), *Towards a Post-Bertalanffy Systemics*, Contemporary Systems Thinking, DOI 10.1007/978-3-319-24391-7_2

correlates of the higher cognitive processes, adopting a non-linear and more systemic approach. Moreover, the integration of neurophysiological studies with brain-imaging studies (fMRI, PET) marked the beginning of a new approach to neuroscience, commonly known as Systems Neuroscience (see, for instance, [23]). Within this context the obtained experimental data allowed us to hypothesize that the mental activity may be the result of a systemic integration of different network activities, not necessarily alternative, based on at least four functional wide range nets (brain networks, BNet) (see, for instance, [3, 12, 16, 18, 19]). These four BNet handle respectively the inner thoughts and memories of the management, evaluation of external stimuli and response behaviors, the level of attention and awareness and the assessment of stimulus meaning. The related neurophysiological underlying phenomena are the expression of a systemic coordination of the activities of a large numbers of neurons. This coordination can be detected in the scalp EEG signal under the form of a phase coherence observed in different frequency bands on more or less extensive areas of the cortex. Some authors attribute to phenomena of phase synchronization at low frequencies also the classical perturbations of EEG waves observed during the first 600 ms after the presentation of a stimulus [8, 10, 14]. These bands seem to correlate with different functions, such as the analysis of the formal characteristics of the stimuli (which occurs locally and which correlate with EEG activity on higher frequency bands), or the processes of long-distance communication, through which meaningful information results in the activation of many different functional areas, more or less extensive, of different networks [12, 20, 24, 25]. Different bands thus play different roles [17] and broader synchronized activations can generate different system phenomena, such as the emergence of conscious thought processes [13, 21]. The mutual interaction between different frequency bands generates the phenomenon of cross-frequency modulation (*CfM*), that not only supports the synchronization and the coordination of the high-frequency activity of areas even very far, but may also represent a generalized mode of systems communication and integration [6].

2 The Cross-Frequency Modulation

As well known, the phenomenon of *CfM* typically concerns the communication between two different brain networks—here called, for simplicity, the sender and the receiver—and consists in the modulation of the amplitude of high-frequency oscillations of the receiver by the phase of low-frequency oscillations of the sender. Such a phenomenon, however, has been sometimes detected also within single networks. The *CfM* appear as characterized by features more or less shared by normal subjects, while pathological subjects (like schizophrenic ones) are characterized by different feature patterns (see, for instance, [2]). Thus, while *CfM* seems to be a robust organizational principle of systemic integration, it could be used also as a diagnostic tool.

Many authors have argued that EEG oscillations in the high-frequency gamma band are local expression analysis of the specific properties of the stimulus, while the low-frequency oscillations are primarily an expression of information processes integration in large functional networks. In this way the integration of the details of the stimulus in more complex psychological entities would be done. At the same time the theoretical data and the simulation data lead us to hypothesize in a similar way that the low-frequency oscillations allow the transmission of information to a wide distance, with higher speeds than the information integrated by local high-frequency activities. This would allow the coordinated analysis on a large scale macro system. These data have an anatomical correspondence in the relation between the white matter myelinated pathways and the gray matter pathways of the brain. The phenomenon of *CfM*, therefore, it is likely to be a key element of the brain integration [9], featuring the learning process [22], the higher cognitive functions as well as the emergence of consciousness [13, 21]. At the same time it could represent also a basic mode in which the information is propagated in scale-free networks, where clusters of the type "small world" are complemented by hub nodes [7]. This type of model finds in the thalamic-cortical-striatal-thalamic circuits their anatomical and functional correspondence [26]. In this perspective it has to be also reminded the heterogeneous structure of the thalamus, in which nuclei, with small-world architecture, are interconnected and separated by areas containing hub neurons (intra-laminar nuclei).

3 An Experimental Study of *CfM*

A preliminary study was conducted on 8 subjects, 4 males + 4 females, right handed, aged between 20 and 51 years old (m.: 32.3 ± 10.3), not suffering of any psychopathology or neurological disorders. The experimental setting consisted in a cognitive task to be completed while EEG signal was recording. EEG signal was recorded at 250 Hz with 256 channels caps by EGI Geodesic equipment. The cognitive task (Cognitive Stress Paradigm—CSP) is a visual paradigm presented on the screen at 60 cm of distance. The CSP is very similar to the Wisconsin Card Sort Test where the geometric pictures presented in the 64 cards are substituted by 64 cartoons. The cartoons concern pictures of landscapes, animals, buildings or plants presented on one card in groups of 1 to 4 elements with four different background colors (light blue, orange, yellow, green). The subject had to match these $4 \times 4 \times 4 = 64$ cards, presented one per time, with four example cards displayed in the top row of the screen. The matching criteria were the background color, the number of the elements inside the picture and the class of picture presented (landscapes, animals, buildings or plants). The computer chose randomly the matching criteria (avoiding to get two equal criteria one after the other) which were blind to the subject and were stable till a specific number of correct responses was reached. This number varied randomly from 8 to 12. The subject had to infer the criteria adopted in that moment by the computer which just gave a feedback to the subject if the response was correct or not. The response had to be given within 5 s. If this time limit was overcome, the response was considered as dropped. The task was stopped when the limit of $64 \times 2 = 128$ presented cards was reached or the subject completed 6 consecutive criteria.

The EEG tracks were processed by means of MatLab and by EEGlab package. The EEG records were reduced from 256 to 71 channels, and cleaned for muscle and EOG artifacts. Then an Independent Component Analysis (ICA) was performed and a dipole best fit computed for each Independent Component (IC). The ICs, with the best fit dipole inside the brain volume and a residual variance of less than 10 %, were considered for further analysis. This ICs pool was segmented in epochs with a time lag ranging from −200 to 800 ms (Time 0 = card presentation start time). Taking into account the subject performance to the previous and to the present trial of CSP, the epochs were labeled in four different groups: *Correct Response after Correct One, Correct Response after an Error, Error Response after Correct, and Error Response after an Error*. The error after the criteria change was excluded by the error responses. The pool of the selected ICs of all subjects was grouped by means of principal component analysis (PCA) and cluster analysis (CA). The statistical tests non executed by MatLab were done by means of the statistical package SPSS.

The observed parameters have been:

1. behavioral parameters describing the performance to the CSP as it is in the original WCST: time to complete task, average trial time, number of completed categories, failures to maintain the set, number of total errors, number of perseverant errors, drops, number of correct responses, total number of presented cards;
2. electrophysiological parameters describing the event related EEG perturbation occurring within 800 ms after the stimulus: the usual ERP waves (N_e, N_1, N_2, P_2, P_{3a}, P_{3b});
3. system integration indices (see [6]): the Cross Frequency Modulation Index ($CfMI$), the Modulation Phase Angle (MPA) and the Resultant Vector Length (MVL) and the residual variance of the associated dipole of the selected ICs (DIP_{rvcm}).

4 Behavioral Data

The mean time to complete the task was 12.0 ± 2.5 min, and completed categories were 4.5 ± 1.7 on average, with a mean time of 1.52 ± 0.14 min/category when the criterion of 6 completed categories was reached. Mean perseverant errors were 7.15 ± 3.3, mean drops 2.18 ± 2.4, mean failures to maintain the set 1.5 ± 1.8. When six categories were completed a mean of 84.7 ± 7.5 cards were shown. Mean response time was 1901 ± 523 ms, with significant differences ($F_{3,166} = 25.33$; $p < 0.001$) between response conditions (*Correct on Correct*: 1489 ± 344 ms; *Error on Correct*: 1816 ± 507 ms, *Correct on Error*: 1996 ± 428 ms; *Error on Error*: 2296 ± 451 ms).

5 Brain Areas Involved in the Event Related Response to CSP

Globally within the 8 subjects considered in this study, 81 ICs were selected and they were aggregated by PCA and CA in 9 clusters: only 4 of these 9 clusters (ACC—*anterior cingulate cortex*, l-OFC—*left orbito frontal cortex*, *visual cortex* and lPOC—*left occipito parietal cortex*) presented significant difference in ERSP in relation to the task performances.

6 Correlation Study About *CfM* Effect Relationship with Neurophysiology and Task Performance

In the whole group of nine clusters the *CfM* parameters were related to the task performance and neurophysiologic parameters by a correlation study (Tables 2.1, 2.2, 2.3, and 2.4).

Table 2.1: Correlation of task performance parameters with *CfM* parameters and the IC-Dipole Residual Variance

$N = 170$	Computed on ICs	Computed on continuous data			Computed on segmented data		
ERP parameter	IC-Dipole Residual variance	Modulation index	Resul. vector length	Mod. phase angle	Modulation index	Resul. vector length	Mod. phase angle
Categories	0.391[a]	–	–	–	–	–	–
Correct responses	−0.189[b]	−0.177[b]	–	–	–	–	–
Cards	−0.261[a]	−0.233[a]	–	0.223[a]	–	–	
Attempts to 1st category	−0.567[a]	–	0.187[b]	−0.228[a]	−0.164[b]	−0.166[b]	–
Drops	−0.554[a]	–	−0.161[b]	–	–	–	
Failures to maintain set	−0.292[a]	−0.206[a]	–	–	−0.166[b]	–	–
Errors	–	–	–	0.230[a]	0.163[b]	–	
Perseverant errors	–	−0.286[a]	−0.185[b]	0.243[a]	–	–	
Response mean time	−0.239[a]	−0.600[a]	–	0.186[b]	−0.262[a]	–	–
Time to complete task	−0.604[a]	–	0.186[b]	−0.262[a]	–	–	

[a] The correlation is significant at the 0.01 level (2-tails)
[b] The correlation is significant at the 0.05 level (2-tails)

7 Modulation Delay (Phase Angle)

The parameter with the highest number of significant correlations has been the *Phase Angle computed on continuous data* (MPA_{cd}), which appears to be higher (this could mean to represent a more delayed *CfM* effect) when *the attempts to*

1st category are lower ($R = -0.228$; $p = 0.003$), the *cards number* is higher ($R = -0.223$; $p = 0.004$), the *time to complete the task* is shorter ($R = -0.262$; $p = 0.001$), but also when *errors* ($R = -0.230$; $p = 0.003$) and *perseverant errors* ($R = -0.243$; $p = 0.001$) are higher. Theses data can be interpreted as indicating that the presence of errors leads to a delay in modulation per modulating frequency, as the number of structures and the length of involved pathways increase.

Table 2.2: Correlation of neurophysiologic parameters with task performance parameters

$N = 170$	Amplitude					
	N_e	N_1	N_2	P_2	P_{3a}	P_{3b}
Categories	–	–	–	–	–	–
Correct responses	0.216[a]	0.256[b]	0.194[a]	–	–	–
Cards	0.161[b]	0.210[a]	–	−0.254[a]	–	–
Drops	–	0.153[b]	–	–	–	–
Failures to maintain set	0.160[b]	0.218[a]	0.185[b]	–	–	–
Perseverant errors	–	0.215[a]	–	0.296[a]	–	–
Global time	–	0.173[b]	–	−0.091[b]	–	–
Errors	–	–	−0.236[a]	–	–	–
Response mean time	–	0.171[b]	–	–	–	–
Cards per condition	0.166[b]	0.226[a]	–	−0.215[a]	−0.156[b]	0.159[b]
Response time per condition	−0.168[b]	−0.178[b]	–	0.158[b]	0. 165[b]	0.175[b]

[a] The correlation is significant at the 0.01 level (2-tails)
[b] The correlation is significant at the 0.05 level (2-tails)

Table 2.3: Correlation of neurophysiologic parameters with task performance parameters

$N = 170$	Latency		
	P_2	P_{3a}	ERP_{1-800} power
Categories	−0.213[a]	–	–
Correct responses	–	–	–
Cards	–	–	–
Drops	–	–	–
Failures to maintain set	–	–	–
Perseverant errors	–	–	–
Global time	–	–	–
Errors	0.160[b]	–	–
Response mean time	–	–	–
Cards per condition	–	–	−0.159[b]
Response time per condition	–	−0.196[b]	

[a] The correlation is significant at the 0.01 level (2-tails)
[b] The correlation is significant at the 0.05 level (2-tails)

Table 2.4: Correlation of neurophysiologic parameters with CfM parameters and the IC-Dipole Residual Variance

$N = 170$	Computed on ICs	Computed on continuous data			Computed on segmented data		
ERP parameter	IC-Dipole residual variance	Modulation index	Resul. vector length	Mod. phase angle	Modulation index	Resul. vector length	Mod. phase angle
N_e amplitude	–	−0.343[a]	–	–	–	–	–
N_1 amplitude	–	−0.317[a]	–	−0.168[b]	−0.375[a]	–	–
N_2 amplitude	–	−0.281[a]	−0.228[a]	−0.187[b]	−0.367[a]	–	–
P_2 amplitude	–	0.179[b]	–	–	0.437[a]	–	–
P_{3a} amplitude	−0.215[a]	–	–	–	0.423[a]	–	–
P_{3b} amplitude	−0.202[a]	–	–	–	0.396[a]	–	–
N_2 latency	0.205[a]	0.166[b]	–	–	0.282[a]	0.307[a]	–
P_2 latency	−0.157[b]	–	–	–	–	–	–
P_{3a} latency	0.173[b]	–	–	0.258[a]	–	–	–
P_{3b} latency	–	–	–	0.220[a]	–	–	–
ERP_{1-800} power	–	–	–	–	0.611[a]	–	–

[a] The correlation is significant at the 0.01 level (2-tails)
[b] The correlation is significant at the 0.05 level (2-tails)

8 Number of Brain Resources Involved (Dipole Residual Variance)

However the *mean cluster residual variance of the matched IC dipoles* (DIP_{rvcm}) has the strongest relationships with the task performance and a high DIP_{rvcm} is correlated with a higher number of completed *categories* ($R = +0.391$; $p < 0.001$), with a lower number of *drops* ($R = -0.554$; $p < 0.001$) and *attempts to complete the 1st category* ($R = -0.567$; $p < 0.001$), with a lower frequency of *failure to maintain the set* ($R = +0.292$; $p < 0.001$), and a shorter *time to complete task* ($R = -0.604$; $p < 0.001$). This means that the single dipole model for those cluster ICs was less probable in spite of a multiple dipole model when the best performances are observed.

The presence of a higher number of *completed categories* was related also to a shorter *latency of* P_2 wave of ERPs ($R = -0.213$; $p = 0.005$) (Table 2.2), which in turn is related to a higher DIP_{rvcm} ($R = -0.157$; $p < 0.040$), (Table 2.3). The presence of a higher DIP_{rvcm} relates with a lower *amplitude* of P_{3a} and P_{3b}, as well as a longer *latency* of N_2 and P_{3a} and a shorter *latency* of P_2. This means that the $lag\,N_2 - P_2$ is shorter and the $lag\,P_2 - P_{3a}$ longer, and can be interpreted as the combined effect of more than one modulating frequency band, a hypothesis that it is coherent with the involvement of many dipoles.

9 Activation of Salience Network (Event Related Modulation Index and N_1-P_2-N_2 Triplet)

The *CfM* Index computed on continuous data ($CfMI_{cd}$) resulted to be higher when the perseverant errors were lower ($R = -0.286$; $p < 0.001$) and the mean response time is shorter ($R = -0.600$; $p < 0.001$). On the other hand the presence of higher frequency of perseverant errors appeared to be linked to a lower value of $CfMI_{cd}$ (Table 2.1), while the same index computed on segmented data ($CfMI_{sd}$) gave no significant results (Table 2.1). The same condition of low frequency of perseverant errors was linked also with a lower response on the N_1 wave ($R = +0.215$; $p = 0.005$) (less deep negativity) and P_2 wave ($R = -0.296$; $p < 0.001$) (less high positivity) on ERP profile (Table 2.2). The presence of a higher $CfMI_{cd}$ and $CfMI_{sd}$ was related to an enhancing effect on N_1, P_2 and N_2 waves (Table 2.3) ($CfMI_{cd}$: N_1, $R = -0.317$, $p < 0.001$; P_2 $R = +0.179$, $p = 0.019$; N_2 $R = -0.281$; $p < 0.001$) ($CfMI_{sd}$: N_1 $R = -0.375$, $p < 0.001$; P_2 $R = +0.437$, $p < 0.001$; N_2 $R = -0.367$; $p < 0.001$). These data are coherent with a synchronization effect of CfM on early EEG events before 300 ms after the card exposure, which leads to an augmented amplitude of the EEG oscillations described by the triplet N_1-P_2-N_2. This triplet usually is correlated with the orienting reaction to novelty or error and to the salience attribution process. We can see also an increase of N_2 latency (the third peak of the triplet) in relation to both $CfMI_{cd}$ ($R = +0.166$, $p < 0.031$) and $CfMI_{sd}$ ($R = +0.282$, $p < 0.001$), a fact which suggests that this synchronization effect may be related to a decrease in the modulating frequency, which in turn can explain the slowing of response reaction when the C wave oscillation (orienting response) increases (Tables 2.2 and 2.3). The higher Pearson R found with parameters computed on segmented data may suggest that this CfM may be actually related to the new stimulus reaction.

10 Error State Management (Non Event Related Modulation Index and N_e)

The N_e wave and early negativity are usually related with the error condition. This circumstance (Table 2.2) appears to be less evident (less negative) when correct responses occur ($R = +0.216$, $p = 0.005$) and it appears to be enhanced (more negative) by a modulating effect of $CfMI_{cd}$ computed on continuous data ($R = -0.343$, $p < 0.001$) (Table 2.4). The non significant correlation with the $CfMI_{cd}$ computed on segmented data (Table 2.4) may suggest a stimulus independent dynamics in such a modulation. It remains here open the question if it can be related to the previous feedback received. Differently from what was found with the N_1-P_2-N_2 triplet, when enhanced the N_e appears to correlate with shorter response time (Table 2.4).

11 Considerations About Experimental Data

When interpreting the previously presented data, we have to take into account the fact that the present study has considered parameters relative to the whole epoch of 5000 ms of inter stimulus time-lag (Cross frequency Modulation and Behavioral Parameters) and parameters relative to the time-lag 0–800 ms after the card onset. Subjects are exposed (during the 5 s of between-cards time lag) to two important stimuli: the card onset, with the consequent orienting, salience and response processing, and the computer reward, occurring 300 ms after the subject response. Responses occurred on average about 1900 ms after the card onset, with a response mean time varying significantly in respect to the context of the previously received reward and to the quality of present response.

Thus our results (especially those linked to neurophysiologic data) are focused mainly on orienting reaction and salience attribution (concerning the salience network activities), and just the early phases only of the response processing. Moreover CfM parameters may represent two kind of conditions: one condition may be linked to the card presentation, grouped by previous response and present performance, and the other may be linked to a general subject attitude toward error risk management, independent from card onset.

Taking into account this framework, the error management as an attitude to expect to deal with an error risk (N_e) seems related more to a stimulus independent attitude then to an event related reactivity. Moreover the presence of an error context seems to correlate with a delay in cross frequency modulation effect alike if, in such a context, more resources and longer pathway could be involved. The involvement of more brain resources as a prerequisite to best performances has a correlate with the increase of residual variance value when best task performances are observed. This attributes to this parameter a meaning supplementing its more obvious one: i.e. in spite of the fact that a lower value of it is considered as a good match for the IC associated dipole, we have, however, that in experimental settings the residual variance can be influenced by the number of dipoles generating that IC. In other words such a hypothesis implies that an EEG IC may be not concerning just only one dipole but it can be produced by the integrated action of many neighboring dipoles. This result is congruent with other experimental evidences, which support the theory that more complex cognitive actions imply the activation and synchronization of many brain structures. Moreover in such conditions, it is possible that the CfM activity may act through different EEG bands (Delta, Theta, Alfa) on many EEG high frequency bands.

Finally the focusing of neurophysiologic parameters on ERPs occurring in the first 500 ms implies that the action of the salience network is more evidenced then the activity of other BNets, as it could be shown by the involvement of the N_1-P_2-N_2 triplet and by the modulation of such activity more likely done by a new stimulus related CfM.

In any case these results confirm the role of CfM in Brain Network activation and modulation. At the same time they highlight the fact that this phenomenon has an inner complexity and modulation in relation to the type of action and to the complexity of processing.

12 Computer Simulations

The existence of a tight interconnection between theory and experiments, which characterizes the systems neuroscience, induced us to resort to computer simulations of neural network activity in order to support the validity of the hypotheses concerning phenomena such as the CfM. In these simulations we used the Integrate-and-Fire neuron models, which are computationally simple enough while, at the same time, being endowed with a good level of plausibility. In logical terms, the dynamical rule of this kind of neurons can be written under the form :

$$if \;\; v(t) < s \;\; then \;\; v(t+\Delta t) = v(t) + [I(t) + a - bv(t)]\Delta t \tag{2.1}$$

$$if \;\; v(t) > s \;\; and \;\; v(t) < A \;\; then \;\; v(t+\Delta t) = A \tag{2.2}$$

$$if \;\; v(t) > s \;\; and \;\; v(t) = A \;\; then \;\; v(t+\Delta t) = c \tag{2.3}$$

Here $v(t)$ is the membrane potential, s a threshold, A the spike amplitude, and c the resting potential. As shown by a number of authors (see [4, 5]) networks made by these neurons exhibit, in presence of a suitable percentage of excitatory and inhibitory connections, globally synchronized oscillations and multistability phenomena.

Our simulations, differently from what can be generally found in literature, deal with networks in which each neuron is characterized by a "personalized" set of parameter values, initially chosen in a random way. The global network behavior, however, is influenced not so much by the inhomogeneity of the neurons, but rather by the nature of their interconnections, chiefly by the presence or absence of long-range interconnections. In this regard, an important feature of the connections is given by the *degree* of each unit, identified with the number of connections (or links) related to the unit itself. We must remind that the *scale-free networks*, already quoted above, are characterized (see [1]) by the fact that the probability distribution of the degree k of their units $P(k)$ depends on k through a law of the form:

$$P(k) = k^{-\gamma} \tag{2.4}$$

These networks, as already mentioned, are characterized by the presence of a small number of *hubs*, that is units with high degree, and a large number of units with small degrees. Some researchers (cfr. [9]) suggested that the phenomenon of *CfM* is a direct consequence of the scale-free character of the biological networks, because the typically large distances between the hubs should favor an information transmission based on large wavelengths (therefore long-ranged) and small frequencies, whose phase should necessarily modulate the short-range high-frequency signals. But the main question is: how the neural networks acquire a scale-free nature?

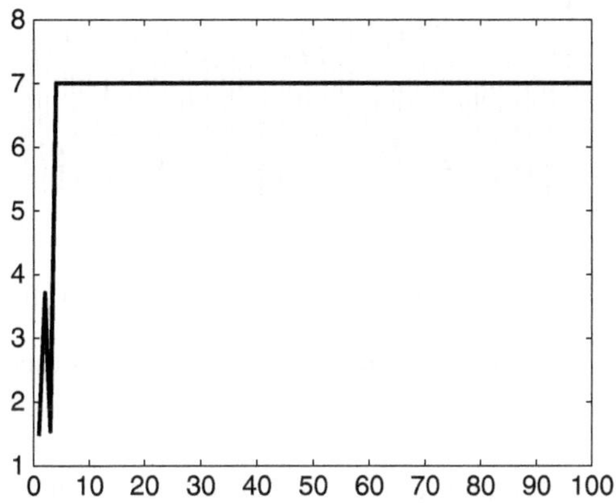

Fig. 2.1: Strong gap junctions (activation vs time)

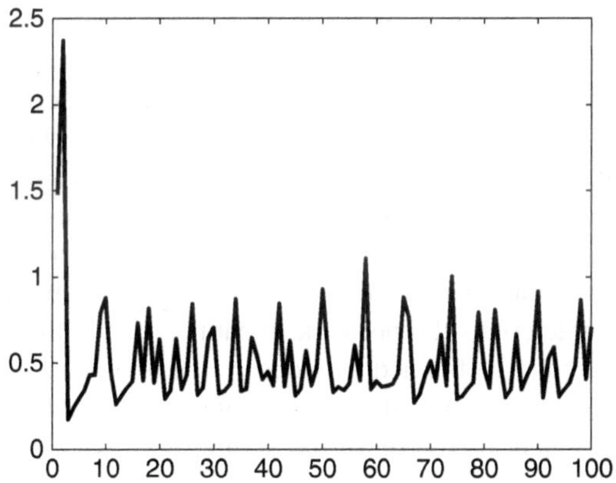

Fig. 2.2: Absence of gap junctions (activation vs time)

To answer we performed a number of simulations including not only synaptic interconnections between neurons but also *gap junctions*. Without entering into technical details (see [15]), we mention that in our simulations, following the method of Lewis and Rinzel [11], the gap junctions act on a neuron through two influences: by directly sending the spikes and by sending a signal dependent on the difference between the membrane potentials of the two involved neurons. Concerning the synaptic efficacies, we postulated an Hebbian law of variation, dependent

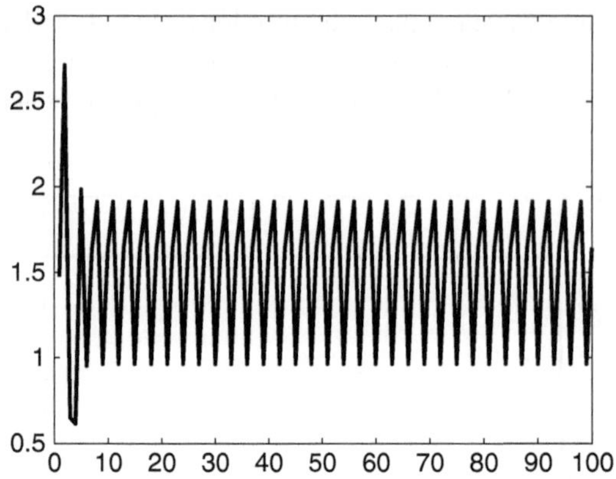

Fig. 2.3: Weak gap junctions (activation vs time)

on spike correlations, including also a spontaneous decay and a saturation effect. The dynamical contributions of the synapses and gap junctions were both weighted, but with different coefficients, so as to allow to investigate the effect on network dynamics of strong, weak, and absent gap junctions. Without showing the results of all simulations performed, we will limit here to present the snapshots of average network activities in three different simulations, all performed on networks of 100 neurons, whose initial connections were given at random.

As it is possible to see from Fig. 2.1, a strong influence of gap junctions leads towards a stable attractor, with a constant value of average activity. However after 100 time steps the network achieved a scale-free structure with $\gamma = 1.28055$. In the case of absence of gap junctions we see from Fig. 2.2 a seemingly noisy average activity, characterized, however, by the presence of different frequency components. In this case the network achieved a strongly scale-free structure with $\gamma = 1.847489$. Finally Fig. 2.3 shows that, in presence of a weak effect of gap junctions, a globally oscillatory state characterizes the network dynamics, while the scale-free features practically disappeared (we obtained a modest $\gamma = 0.3550416$).

These results seem to evidence that the gap junctions could act as modulators of the network structure and, therefore, of the nature of global activity it can support. It remains to be found what are the physiological processes implementing such a modulation activity. Answering this question would be equivalent to understand in what circumstances we could expect a normal operation of CfM.

13 Conclusions

In principle, it would be possible that a network, with synaptic weights varying as a function of the inner activity according to an Hebbian-like law, could reach a scale-free structure, if supported by a suitable influence of gap junctions. This would open the way to a conception of the CfM as an almost universal systemic feature of the normal operation of most systems (not only of neural networks). If this would be the case, then this model could be exported also to more complex systems such as social networks, in which micro-social units with interpersonal high density relationships similar to "small world" architectures can be integrated in macro-social structures in which the small world elements are mutually functionally integrated by individuals with the ability to "hubbing" through low frequency collective activities.

References

1. Albert, R., & Barábasi, A.-L. (2002). Statistical mechanics of complex networks. *Reviews of Modern Physics, 74*, 47–97.
2. Allen, E. A., Liu, J., Kiehl, K. A., Gelemter, J., Pearlson G. D., Perrone-Bizzozero, N. I., & Calhoun, V. D. (2011). Components of cross-frequency modulation in health and disease. *Frontiers in Systems Neurosciences, 5*, 1–16.
3. Bressler, S. L., & Menon, V. (2010). Large-scale brain networks in cognition: Emerging methods and principles. *Trends in Cognitive Sciences, 14*, 277–290.
4. Brunel, N. (2000). Dynamics of sparsely connected networks of excitatory and inhibitory spiking neurons. *Journal of Computational Neuroscience, 8*, 183–208.
5. Brunel, N., & Hakim, V. (2008). Sparsely synchronized neuronal oscillations. *Chaos, 18*, 015113.
6. Canolty, R. T., & Knight, R. T. (2010). The functional role of cross-frequency coupling. *Trends in Cognitive Sciences, 14*(11), 506–515.
7. Da Rocha, E. L., & da Cunha, C. R. (2011). The transition from fracton to phonon states in a Sierpinski triangle lattice. *Chaos, Solitons & Fractals, 44*, 241–247.
8. Gruber, W. R., Klimesch, W., Sauseng, P., & Doppelmayr, M. (2005). Alpha phase synchronization predicts P1 and N1 latency and amplitude size. *Cerebral Cortex, 15*, 371–377.
9. He, B. J., Zempel, J. M., Snyder, A. Z., & Raichle, M. E. (2010). The temporal structures and functional significance of scale-free brain activity. *Neuron, 66*, 353–369.
10. Klimesch, W., Schackb, B., Schabusa, M., Doppelmayr, M., Gruber, W., & Sauseng, P. (2004). Phase-locked alpha and theta oscillations generate the P1-N1 complex and are related to memory performance. *Cognitive Brain Research, 19*, 302–316.
11. Lewis, T. J., & Rinzel, J. (2003). Dynamics of spiking neurons connected by both inhibitory and electrical coupling. *Journal of Computational Neuroscience, 14*, 283–309.
12. Meehan, T. P., & Bressler, S. L. (2012). Neurocognitive networks: Findings, models, and theory. *Neuroscience and Biobehavioral Reviews, 36*, 2232–2247.
13. Melloni, L., Molina, C., Peña, M., Torres, D., Singer, W., & Rodriguez, E. (2007). Synchronization of neural activity across cortical areas correlates with conscious perception. *The Journal of Neuroscience, 27*(11), 2858–2865
14. Muller, V., & Anokhin, A. P. (2012). Neural synchrony during response production and inhibition. *PLoS One, 7*(6), e38931.
15. Ostojic, S., Brunel, N. & Hakim, V. (2009). Synchronization properties of networks of electrically coupled neurons in the presence of noise and heterogeneities. *Journal of Computational Neuroscience, 26*, 369–392.
16. Park, H.-J., & Friston, K. (2013). Structural and functional brain networks: From connections to cognition. *Science, 342*, 1238411.
17. Peña, M., & Melloni, L. (2012). Brain oscillations during spoken sentence processing. *Journal of Cognitive Neuroscience, 24*(5), 1149–1164.
18. Raichle, M. E. (2011). The restless brain. *Brain Connectivity, 1*(1), 3–12.
19. Sporns, O. (2010). *Networks of the brain*. MIT Press: Cambridge, MA.

20. Steinke, G. K., & Galan, R. F. (2011). Brain rhythms reveal a hierarchical network organization. *PLoS Computational Biology, 7*(10), e1002207.
21. Tononi, G. (2010). Information integration: Its relevance to brain function and consciousness. *Archives Italiennes de Biologie, 148*, 299–322.
22. Tort, A. B., Komorowski, R. W., Manns, J. R., Kopell, N. J., & Eichenbaum, H. (2009). Theta-gamma coupling increases during the learning of item-context associations. *Proceedings of the National Academy of Sciences of the United States of America, 106*, 20942–20947.
23. van Hemmen, J. L., & Sejnowski, T. J. (Eds.). (2006). *23 problems in systems neuroscience*. Oxford University Press: New York.
24. von Stein, A., & Sarnthein, J. (2000). Different frequencies for different scales of cortical integration: From local gamma to long range alpha theta synchronization. *International Journal of Psychophysiology, 38*, 301–313.
25. Yeung, N., Bogacz, R., Holroyd, C. B., Nieuwenhuis, S., & Cohen, J. D. (2007). Theta phase resetting and the error-related negativity. *Psychophysiology, 44*, 39–49.
26. Zamora-Lopez, G., Xhou, C., & Kurths, L. (2011). Exploring brain function from anatomical connectivity. *Frontiers in Neurosciences, 5*(83), e2011.00083.

Chapter 3
Testing Different Learning Strategies on a Simple Connectionist Model of Numerical Fact Retrieval

Simone Pinna and Giorgio Fumera

One of the main topics of *Systems theory* is the development of *nonlinear* models for the simulation of complex interactions between a system and its environmental conditions.[1] In this paper we present a connectionist model of simple arithmetic in order to investigate if different learning strategies (seen as the *external* conditions of that model) may influence its performances independently of systemic properties.

1 Number Representation

In the field of cognitive arithmetic, connectionist systems have been used to model basic numerical skills such as number comparison, subitizing, counting, and to simulate mental arithmetic. The number encoding schemes, which should reflect relevant features of the representation of numerical magnitude in human beings, have a crucial role in these works.

[1]For a review of current topics and tools of Complex Systems Theory [29] see chapter 0 of Bar-Yam [5].

S. Pinna (✉)
Alophis – Applied Logic Language Philosophy and History of Science, University of Cagliari, Via is Mirrionis 1, 09123 Cagliari, Italy
e-mail: simonepinna@hotmail.it

G. Fumera
Department of Electrical and Electronic Engineering, Piazza d'Armi S.n.c., 09123 Cagliari, Italy
e-mail: fumera.diee@unica.it

G. Minati et al. (eds.), *Towards a Post-Bertalanffy Systemics*, Contemporary Systems Thinking, DOI 10.1007/978-3-319-24391-7_3

1.1 Number Encoding Schemes

Four different schemes for encoding numerical representation, typically thought to include magnitude information, have been used in connectionist models [33]:

1. **"Barcode" magnitude representation** [2, 3]. Numbers are encoded as sets of adjacent active units of an ordered set of nodes , where each node is labeled with a particular number. In this scheme, e.g., the number 7 is encoded via the activation of the nodes labeled "6", "7" and "8".
2. **Compressed number line** [10]. Each number is represented by a pattern of activation of the input neurons, where a neuron set to 1 is surrounded by noisy neurons activated according to a gaussian with fixed variance. In this scheme, the number series is generated by following a logarithmic scale, in such a way that representations of larger numbers share more active neurons than representations of smaller numbers. This method of representing numbers has the purpose of mirroring empirical evidences which indicate that smaller numerosities are more easily discriminated among them than larger ones.
3. **Number line with scalar variability** [10, 14]. This scheme differs from the compressed number line for the number series is held in a linear fashion. For any number n the total activation is constant while the variance is proportional to n itself. Interestingly, Dehaene showed that encoding numbers this way led to the same net behavior as using the compressed number line.
4. **Numerosity coding structure** [9, 31, 32]. Each number n is represented by the activation of n units of a linear set of neurons, where the activation of the first, leftmost neuron represents number 1, the activation of the first and the second leftmost neurons represents number 2, and so on.

In the preceding sketch, the schemes at points 2 and 3 are thought to be consistent with the mental number-line theory [11, 12], according to which successive numbers are represented analogically as increasingly closer points of a straight line. On the other way, the numerosity coding structure encodes numbers digitally, so that bigger numbers include smaller numbers.

1.2 Numerosity Coding: Simulations

The numerosity coding structure has been employed in neural network simulations of several numerical skills, such as number comparison [32], and simple arithmetic operations [25–27].
In a series of interesting simulations, Stoianov et al. [25] tested the computational properties of the aforementioned four different ways of number encoding on an *associative memory* model, namely a Boltzmann Machine [1] trained with the contrastive divergence algorithm [28] in order to learn to carry out single-digit additions (SDAs). After the training phase, the associative network was able to retrieve complete arithmetic facts, namely both arguments and the result of each SDA, starting

from incomplete patterns. The net was tested on this main task by employing four different number encoding schemes: a *symbolic code*, where each number is represented by the activation of a dedicated node, and the schemes 2–4 seen in the previous paragraph. Interestingly, when using the numerosity structure, the network was able to retrieve correct outputs easier—i.e., after less presentations of the complete training set (epochs).
The crucial experiment where the numerosity coding revealed its better capacity to reproduce real cognitive phenomena was made on the analysis of the so called *problem-size effect* [4, 17, 19, 22, 30].[2] Only when using the numerosity structure, indeed, the reaction time distribution of the net that was in line with human data.
In a successive work, Stoianov et al. [26] expanded the net architecture with a symbolic component. This further step allowed to investigate the relations between semantic and symbolic representations in the development of simple arithmetic skills [20], and revealed a dynamic in which semantic representations seem to have a predominant role for operation performances, i.e. the task was solved first on the semantic side of the net, and then the correspondent symbolic counterpart was activated.

The simulations presented above were expressly designed in order to test structural properties of the connectionist systems used, e.g. to test the cognitive plausibility of number encoding schemes or to assess the mutual role of various system components. In this article we propose a different, *algorithmic* approach to the analysis of early arithmetic skills and use a connectionist system to try the effects of different *learning strategies* (simulated by the manipulation of training sessions) on the behavior of the model.

2 Finger Counting and Arithmetic Development

The use of fingers for counting plays an acknowledged role in the development of early arithmetic skills [6–9, 13, 18].
A recent research [24] brings evidences that finger gnosia, i.e. the correct representation of fingers, is associated on the one side to a greater probability of finger-use in computation and, on the other side, to better arithmetical performance in 5–7 years old children. Given that relations between poor finger gnosia and poor arithmetical skills have also been found [21], it should be interesting to inspect the cognitive mechanism on which this phenomenon is grounded.
The main question on which we will focus our attention is, hence, the following: *how does finger use in counting routines affects learning of basic arithmetical facts?*
We propose a twofold method to face this question. First, we will analyze finger-using counting strategies from an algorithmic stance. Through this analysis it could be possible to inspect the relevant operations and the implicit knowledge neces-

[2] This effect consists of an increase in reaction time and error rate in response to arithmetic problems with larger solutions (i.e., solving $7+8$ takes longer and is more error-prone than solving $4+3$).

sary to perform that specific counting routine. Second, the algorithmic features of the counting routine will be simulated on a feed-forward back-propagation neural network designed to study the learning and retrieval of a set of basic number facts—namely, the results of SDAs. The net is given different training-sets in order to simulate different ways of acquisition of SDA results, one of which is modeled on the finger-counting routine.

2.1 Algorithmic Features of the Finger Counting Routine

When children learn basic arithmetic concepts, they spontaneously develop counting routines that normally imply the use of objects, *abaci*, fingers and so on [15]. The most elementary counting routine is the so-called *counting-all* strategy, which consists on counting out the first set, then the second, and lastly the combination of the two sets.
A slightly more advanced counting routine is the *counting-on* strategy, commonly carried out with the use of fingers, which is performed by counting out the value of the second addend from the value of the first up to the result.
If we focus on the basic knowledge children have to possess in order to carry out the counting-on strategy for SDAs, it seems obvious that they need at least to master the successor function in the domain $\{0,\ldots,17\}$.[3]
Once a child learns to use the counting-on strategy, we could conjecture that she would apply it to SDAs presented randomly rather than according to a given order (e.g., first all the 1-digit sums $1+n$, then $2+n$, and so on).
These considerations will be useful for the manipulation of training procedures that will be tested on a feed-forward network designed to simulate retrieval of SDA results.

2.2 Net Description and Experiments

We conducted experiments using a feed-forward, fully-connected multi-layer perceptron neural network.
The input layer encodes two single-digit numbers x_1 and $x_2 \in \{0,\ldots,9\}$. The network was trained such that the output layer encodes the sum $x_1+x_2 \in \{0,\ldots,18\}$. The numbers are encoded using the numerosity coding structure; therefore, both the input and output layers are made up of 18 neurons (see Fig. 3.1).
One hidden layer of five neurons was used.

[3] For the larger SDA is $9+9$. Pinna [23] proposes a formal description of the counting-on algorithm through a Bidimensional Turing machine [16].

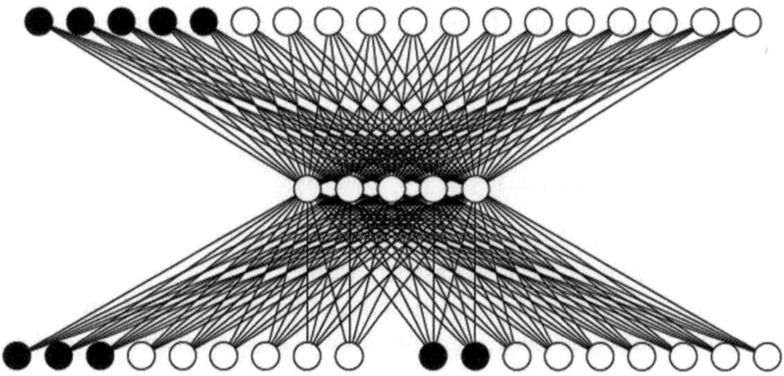

Fig. 3.1: Structure of the neural network used for the experiments. In this example, the input layer encodes the numbers $x_1 = 3$ and $x_2 = 2$, the output layer encodes their sum 5

The activation a of each neuron equals the weighted sum of its inputs, plus a bias term. The following activation function was used: $(1+exp(-ka))^{-1}$, with $k = 0.5$ for hidden neurons, and $k = 1.0$ for output neurons.
We used the standard back-propagation as the learning algorithm, with a learning rate of 0.1.

We carried out four experiments, using different strategies for network training. Strategy (a) simulates a rote learning strategy, as if a subject were using an "addition table" to learn all the SDAs.
Strategy (d) is designed to reflect the acquisition of results given by the use of a counting-on strategy as that used in a finger counting routine (see Sect. 2.1) and consists of two steps: first, the network is trained on a finite domain of the successor function; second, it is trained on all the SDAs in random order.
Strategies (b) and (c) are intermediate between the previously described ones and are used as control.

Here follows a detailed description of the experiments:

(a) The training set is made up of all possible examples (sums) that are processed at each epoch of the back-propagation algorithm in the following order:

$$0+0,0+1,0+2,\ldots,0+9,$$
$$1+0,1+1,1+2,\ldots,1+9,$$
$$\ldots$$
$$9+0,9+1,9+2,\ldots,9+9.$$

(b) The training set is made up of all possible examples in random order; they are processed in the same (random) order at each epoch.

(*c*) The network is first trained on the 20 sums of the form $x_1 + 1$ and $1 + x_2$, in the following ordering, until no errors is made on them:

$$0+1, 1+1, 2+1, \ldots, 9+1,$$

$$1+0, 1+1, 1+2, \ldots, 1+9;$$

then, starting from the learned weights, the network is trained on all the examples, processed in the same order as in case (*a*) above.

(*d*) The network is first trained as in case (*c*) above, then it is trained on all the examples in random order (the same, random order was used on all the epochs).

We evaluated the error rate on all the 100 examples. Then, to verify if the net could reproduce the problem size effect, we separately evaluated the error rate on "small-size problems" (i.e., the 49 sums $x_1 + x_2$, with $x_1, x_2 \leq 6$) and on "large-size problems" (the remaining 51 sums, with x_1 or $x_2 > 6$).

The above experiments were run for ten times, starting from different initial weights (randomly chosen) of the network connections. The results are reported in terms of the average error rate over the ten runs. Results are shown in Figs. 3.2 and 3.3.

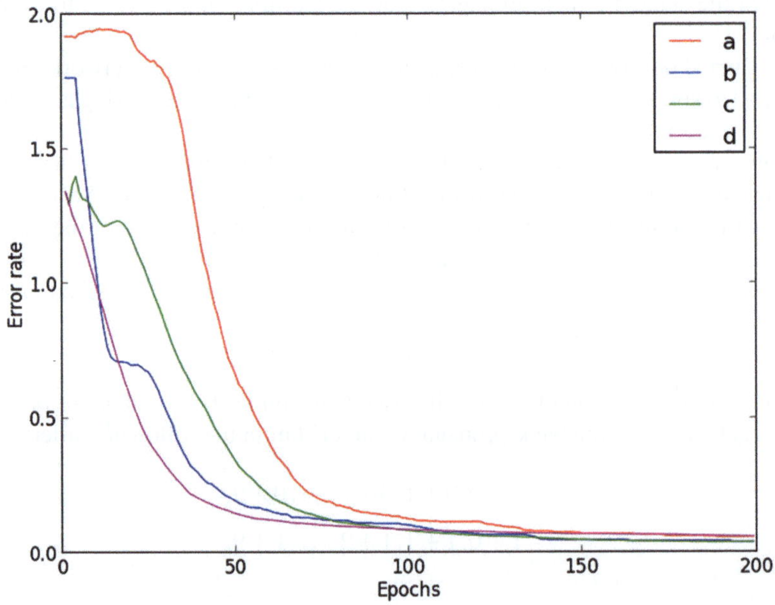

Fig. 3.2: Comparison of the learning curves (error rate per epoch) correspondent to the different training conditions (*a*)–(*d*)

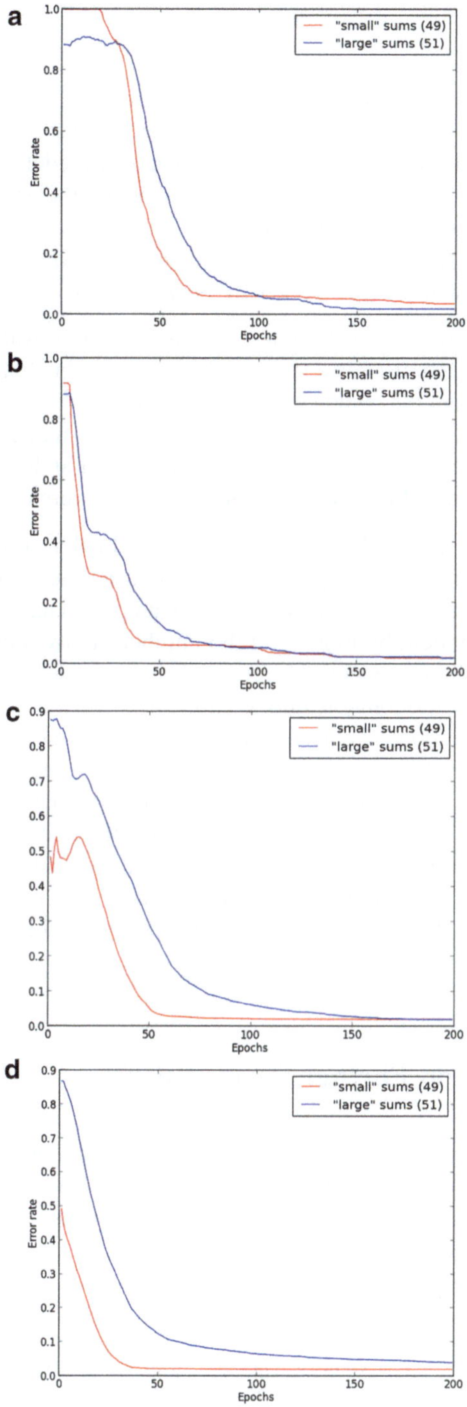

Fig. 3.3: Problem size effect evaluated as error rate per epoch respectively, in training condition (**a**)–(**d**)

2.3 Discussion

We tested different strategies for learning SDA results on a feed-forward neural net to check if we could find any effect linked to the learning strategy adopted. First, we evaluated the possible advantages of using the strategy modeled on the finger counting routine (strategy *d*). Moreover, we tested the net on the problem size effect[4] for each of the learning strategies simulated.
The adoption of a feed-forward neural architecture, which is non-standard for numerical skill simulations (see Sect. 1.2), is justified by two reasons:

1. Given that we were specifically interested on result retrieval of SDAs, the use of this neural architecture allows to keep distinct the arguments of the function in the input layer from the results in the output layer.
2. Our simulation was aimed to test learning strategies independently of systemic properties of the network; hence we needed a neural architecture whose properties were not, alone, able to simulate the effects we were interested on.

Learning strategy (*d*) leads the net to a faster reduction of the number of errors from the $20th$ epoch on (see Fig. 3.2). This could be due to the fact that, in (*d*), the net is already trained on a subset of the examples, i.e., on all the sums $n+1$ and $1+n$. This is evident by looking at the number of errors committed by the net at the very first training epochs. Moreover, learning strategy (*a*) leads the net to concentrate probably too long on similar cases.
Given these considerations, we needed a way to check some effect of learning strategy which would not be trivially ascribed to systemic properties or to previously learned information. So we tested the net in order to verify whether and in what training conditions it can reproduce the problem size effect. The results here were very interesting. Learning strategy (*d*), indeed, leads the net to reproduce the problem size effect in a very smooth fashion on the entire plot (namely, at least until the $200th$ epoch), while the in the "rote learning" strategy and in the two intermediate ones this effect is not verified (see Fig. 3.3).

3 Conclusions and Further Developments

We tried to verify the effect of different learning strategy, one of which is in accordance with the finger-counting routine, by using a feed-forward neural net in order to simulate the acquisition of a definite set of number facts, namely the set of all the possible 1-digit addition results.

---

[4] With regard to the above described net it is possible to ascertain only one side of the problem-size effect, namely if, during the training phase, the net is more error prone on training set cases where the solution is larger. Temporal features, indeed, cannot be simulated in a standard feed-forward net.

The behavior of the net has been observed in different training conditions. The training condition suggested by the algorithmic analysis of the finger-counting routine is indeed more convenient for a faster reduction of net's errors during the training phase.

However, this fact alone is insufficient to exclude a trivial explanation of net's behavior based on its structural properties. Then, to verify the cognitive plausibility of the model, this has been tested on the problem-size effect. The results showed that the training condition modeled on the finger-counting strategy leads not only to a faster reduction of errors during the training phase, but also provides to the net the ability to reproduce a well-known cognitive effect.

Further developments may include the manipulation of two main variables:

1. The number encoding scheme. If the effects of the learning strategy do not depend on network properties, they should be verified even if numbers are encoded differently (namely, by using any of the four main schemes seen in Sect. 1.1).
2. The neural architecture. The effects of learning strategies should be reproduced whatever neural architecture is adopted.

References

1. Ackley, H., Hinton, E., & Sejnowski, J. (1985). A learning algorithm for Boltzmann machines. *Cognitive Science, 9*, 147–169.
2. Anderson, J. (1998). Learning arithmetic with a neural network. In D. Scarborough & S. Stenberg (Eds.), *An invitation to cognitive science, volume 4: Methods, models and conceptual issues* (pp. 255–300). Cambridge, MA: MIT Press.
3. Anderson, J. (2002). Hybrid computation with an attractor neural network. In *First IEEE International Conference on Cognitive Informatics, 2002. Proceedings* (pp. 3–12). doi:10.1109/COGINF.20021039275
4. Ashcraft, M. (1992). Cognitive arithmetic: A review of data and theory. *Cognition, 44*, 75–106.
5. Bar-Yam, Y. (1997). *Dynamics of complex systems* (Vol. 213). Reading, MA: Addison-Wesley.
6. Butterworth, B. (1999). A head for figures. *Science, 284*, 928–929.
7. Butterworth, B. (2000). *The mathematical brain*. New York: Papermac.
8. Butterworth, B. (2005). The development of arithmetical abilities. *Journal of Child Psychology and Psychiatry, 46*, 3–18.
9. Butterworth, B. (2010). Foundational numerical capacities and the origins of dyscalculia. *Trends in Cognitive Sciences, 14*, 534–541.
10. Dehaene, S. (2001). Subtracting pigeons: Logarithmic or linear? *Psychological Science, 12*, 244–246.
11. Dehaene, S. (2003). The neural basis of the weber-fechner law: A logarithmic mental number line. *Trends in Cognitive Sciences, 7*(4), 145–147.
12. Dehaene, S. (2011). *The number sense: How the mind creates mathematics* (revised and updated edition). Oxford: Oxford University Press.
13. Fuson, K. (1988). *Children's counting and concepts of number*. New York: Springer.
14. Gallistel, C., & Gelman, R. (1992). Preverbal and verbal counting and computation. *Cognition, 44*(1), 43–74.
15. Gelman, R., & Gallistel, C. (1978). *The child's understanding of number*. Cambridge, MA: Harvard University Press.
16. Giunti, M. (2009). Bidimensional Turing machines as Galilean models of human computation. In G. Minati, M. Abram, & E. Pessa (Eds.), *Processes of emergence of systems and systemic properties*. Singapore: World Scientific.

17. Groen, G., & Parkman, J. (1972). A chronometric analysis of simple addition. *Psychological Review, 79*, 329–343.
18. Lakoff, G., & Núñez, R. (2000). *Where mathematics comes from: How the embodied mind brings mathematics into being*. New York: Basic Books.
19. Le Fevre, J., Sadesky, G., & Bisanz, J. (1996). Selection of procedures in mental addition: Reassessing the problem size effect in adults. *Journal of Experimental Psychology: Learning, Memory, and Cognition, 22*, 216–230.
20. Mc Closkey, M., Aliminosa, D., & Sokol, S. M. (1991). Facts, rules and procedures in normal calculation: Evidence from multiple single-patient studies of impaired arithmetic fact retrieval. *Brain and Cognition, 17*(2), 154–203.
21. Noël, M. (2005). Finger gnosia: A predictor of numerical abilities in children? *Child Neuropsychology, 11*, 413–430.
22. Núñez-Peña, M. (2008). Effects of training on the arithmetic problem-size effect: An event-related potential study. *Experimental Brain Research, 190*, 105–110.
23. Pinna, S. (2014). *Extended cognition, dynamics, and algorithm: A Turing machine based approach to the study of algorithmic skills*. Doctoral thesis, Università di Cagliari.
24. Reeve, R., & Humberstone, J. (2011). Five-to-7-year-olds' finger gnosia and calculation abilities. *Frontiers in Psychology, 2*, article 359.
25. Stoianov, I., Zorzi, M., Becker, S., & Umiltà, C. (2002). Associative arithmetic with Boltzmann machines: The role of number representations. In *Artificial Neural Networks – ICANN 2002* (pp. 277–283). New York: Springer.
26. Stoianov, I., Zorzi, M., & Umiltà, C. (2003). A connectionist model of simple mental arithmetic. In *Proceedings of EuroCogSci03* (pp. 313–318).
27. Stoianov, I., Zorzi, M., & Umiltà, C. (2004). The role of semantic and symbolic representations in arithmetic processing: Insights from simulated dyscalculia in a connectionist model. *Cortex, 40*(1), 194–196.
28. Welling, M., & Hinton, G. E. (2002). A new learning algorithm for mean field Boltzmann machines. In *Artificial Neural Networks – ICANN 2002* (pp. 351–357). New York: Springer.
29. Wolfram, S. (1988). Complex systems theory. In D. Pines (Ed.), *Emerging Syntheses in Science. Proceedings of the Founding Workshops of the Santa Fe Institute*. Westview Press: Santa Fe, NM.
30. Zbrodov, N., & Logan, G. (2005). What everyone finds: The problem-size effect. In J. Campbell (Ed.), *Handbook of mathematical cognition* (pp. 331–346). Psychology Press: New York.
31. Zorzi, M., & Butterworth, B. (1997). On the representation of number concepts. In M. Shafto & P. Langley (Eds.), *Ninth Annual Conference of the Cognitive Science Society*. Mahwah, NJ: LEA.
32. Zorzi, M., & Butterworth B. (1999). A computational model of number comparison. In M.H.S. Stoness (Ed.), *Proceedings of the Twenty First Annual Conference of the Cognitive Science Society*. Mahwah, NJ: Erlbaum
33. Zorzi, M., Stoianov, I., & Umiltà, C. (2005). Computational modelling of numerical cognition. In J. Campbell (Ed.), *Handbook of mathematical cognition* (pp. 67–84). Hove: Psychology Press.

Chapter 4
Dynamical Systems and Automata

Mario R. Abram and Umberto Di Caprio

1 Introduction

Dynamical systems in their mathematical representation are a powerful tool for describing the structure and the evolution of systems. In particular, state space is a geometric representation of the many possible evolutions of the system's state. This representation is particularly effective and useful in describing the dynamics of the state for nonlinear systems, by means of which a qualitative description of the system becomes possible.

We assume valid the principle of conservation of energy. Instead the geometric space represents the evolution of the system state variables. The dimension of state space is given by the number of state variables. It gives a global, synthetic representation of the possible evolution of the system state. "Global" means that it gives information about all the system; "synthetic" means that it gives effective, appreciable, valuable information about the system.

Some questions arise and orientate the development of our analysis. Can we extract the essential information about the system from state space? Can we use such an information to describe and characterize the system? Can we formalize a process for describing qualitatively the evolution of the system? Can we find a simplified description of the system that gives us a synthetic but complete description of the system evolution? May this process for information extraction be formalized? Can it become the basis for an operating methodology?

Often these questions arise in engineering when we need to build a simplified model for evaluating, setting and planning the design activities.

M.R. Abram (✉)
AIRS (Associazione Italiana per la Ricerca sui Sistemi), Milano, Italy
e-mail: mario.abram@alice.it

U. Di Caprio
ASDE (Associazione Dirigenti Enel), Roma, Italy
e-mail: umbertodicaprio@tiscali.it

G. Minati et al. (eds.), *Towards a Post-Bertalanffy Systemics*, Contemporary Systems Thinking, DOI 10.1007/978-3-319-24391-7_4

In this paper, after recalling some properties of dynamical systems (Section 2), we consider the partitions of the state space (Section 3) and we define the automaton associated to a partition (Section 4). These concepts are illustrated by the example of classical pendulum, considering the conservative and dissipative cases (Section 5). Finally, some remarks and final conclusions recall and suggest possible future researches and developments (Section 6 and 7).

2 Dynamical Systems

Dynamical Systems are usually described by a system of n ordinary differential equations and the state vectors are located into a geometrical space defined by a state variable associated to each coordinate axis [1, 7, 8].

So, for example, a pendulum is described by a second order system ($n = 2$) of two nonlinear ordinary differential equations; the state variables are the angular position and the angular speed.

State Space is the geometric locus of the trajectories of the possible evolutions of the systems state variables. The dimension of state space is given by the number of state variables. Historically we heard the terms phase space, phase portrait but in general, especially for control applications and modeling, the expression "state space" became more usual. In particular the phase portrait shows how the system can evolve in time starting from each initial condition point in state space.

Then the state space representation can be seen as a qualitative description of the system. By means of qualitative study of the system, information becomes available about evolution in time, critical points, stability, etc. [2, 3, 7–9]. State Space plus Time coordinate gives the Movement Space. It gives the trajectory of the system versus time.

3 State Space Partition

We consider a partition of state space based on criteria giving a subdivision of the state space into volumes. The state space can be partitioned in different ways. We consider mainly two distinct ways.

- *Geometric Partition.* State Space partition is obtained by means of an orthogonal grid built as a set of straight lines orthogonal to the axes of the state space. The grid defines the volumes of each partition.
- *Physical Partition.* An alternative State Space Partition (we can call it physical) can be defined considering the trajectories in state space. State Space partition is obtained by means of manifolds of state space description.

A partition subdivides the state space into cells and it is interesting to find e simplified way to describe the relations between the cells and the borders separating them. The conditions useful to describe these relations can change depending on the chosen partition, but some aspects can be collected and syntetized into very basic properties. These mathematical properties can be connected each other forming a network that, as we will see, can assume quite easily the structure of an automaton.

4 Associated Automata

The concept of Automaton was developed under the need to describe discrete systems and enriched itself by many contributions. So different automata definitions are available and we can recall some examples as finite automata, Mealy and Moore automata, Petri nets, etc. They are continuously evolving and new results become often available [6]. Automata and all the modeling and simulation components are a fundamental tool in designing and testing control systems [4]; but they become also an alternative description of physical systems sinthetizing information into essential concepts [10].

Given a partition of the state space, it is possible to associate two different states that describe intuitively two different situations: the position in the state space and the movement between two positions in state space. Then it is convenient to distinguish between *Fundamental States* and *Transition States*:

- *Fundamental State*. It occurs when the status of the system is part of a specific cell generated by the partition. This state is unique because it is associated to a single cell. The fundamental state may have many inputs and many outputs.
- *Transition State*. It occurs when the status of the system changes (moves) from a cell to another cell of the partition. This state defines the transition between two cells. Consequently there is a number of these states proportional to the number of contiguous cells. They have only one input and only one output.

Then we can go into a fundamental state from many transition states and from a fundamental state we can go into many transition states; Indeed we go into a transition state only from a fundamental state and from a transition state we can go only into a fundamental state.

Connecting all the transition and fundamental states of a partition, the resulting network is an automaton that we call *Associated Automaton* to the dynamical system.

Given a dynamical system, the previous concepts can be collected into a procedure composed by the following steps:

1. Definition of the state space and preparation of the state portrait.
2. Choice of the partition (geometric or physical or other …), taking into account the structure of the state portrait.
3. A fundamental state is associated to each cell of the partition.
4. Two transition states are associated to each border between two cells, one for each crossing direction.
5. By connecting the fundamental states by means of the couples of transition states one obtain an automaton.

6. Then an automaton is defined from the structure of state space partition.

The states of the associated automaton are defined substantially by the topology of state space description, and the number of involved states is given by $N_F + N_T$ where:

N_F is the number of fundamental states (it is related to the number of cells between the manifolds involved).

N_T is the number of the couples of transition states (it is defined by the number of crossing sections between the manifolds).

5 An Example

The classical pendulum is an useful example to illustrate the basic concepts of this representation. In figure 4.1 the potential function and the phase space of the classical pendulum are represented; the system is conservative and the represented curves are the trajectories generated by the pendulum mass moving with a total energy that is constant, when different values of the energy are assigned. The represented state space is formed by the trajectories of the mass at different energy levels. In particular the trajectory connecting critical unstable points corresponds to a specific value of energy.

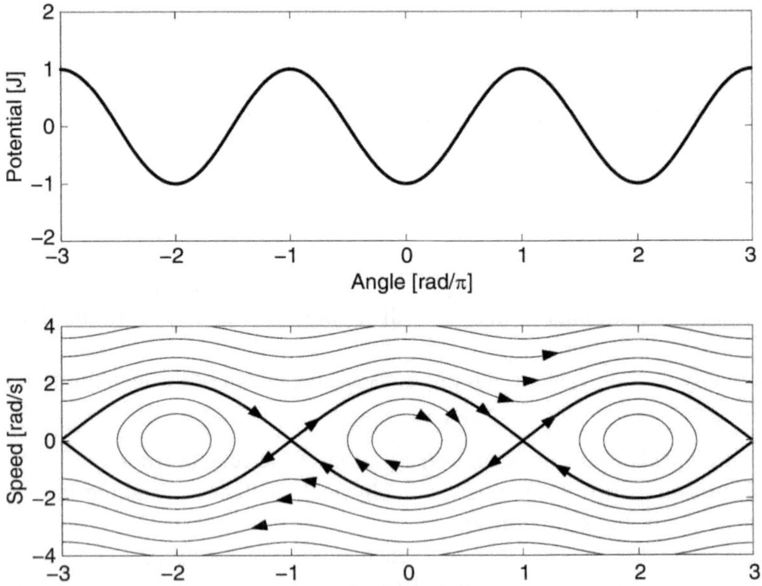

Fig. 4.1: Classical conservative pendulum. Potential function of the system and the State Space

In order to test the application of the previously discussed ideas we will apply the procedure of Sec. 4 to the test case of classical pendulum, considering some important implications due to conservative and non conservative cases. We will apply two partitions, a so called geometric one and a physical one, to the portion of state space defined by the range of validity $-\pi \leq \theta \leq \pi$.

The geometric partition is defined by means of two orthogonal sheaves of straight lines defined by $\theta = 0$ and $\theta = \pm\pi$ for vertical lines and $\dot{\theta} = 0$ and $\dot{\theta} = \pm k$ for horizontal lines (Figure 4.2(left)). The state space is subdivided into 8 cells, then 8 fundamental states (A,B,C,D,E,F,G,H) are associated to the cells. These 8 fundamental states are connected by 36 transition states (Figure 4.3(left)) considering the extensions on the left and the right of the state space ($N_F = 8$ and $N_T = 36$).

In an analogous way another partition (physical) is defined by two straight lines ($\theta = -\pi$ and $\theta = \pi$) and the energy curves connecting the two critical unstable points $(-\pi, 0)$ and $(\pi, 0)$ (Figure 4.2(right)). In this way we created three cells, then again 3 fundamental states (R – Right rotation; L – Left rotation; P – Oscillating Pendulum) are associated to the cells. These 3 fundamental states are connected by 12 transition states (Figure 4.3(right)) considering the extensions on the left and the right of the state space ($N_F = 3$ and $N_T = 12$).

Given two different partitions of the same state space, the resulting associated automata are very different as the number of states shows: ($N_F = 8$ and $N_T = 36$) for geometric partition; ($N_F = 3$ and $N_T = 12$) for physical partition.

The trajectories of a conservative system coincide with the energy curves that define the state space (Figure 4.4(left)). Instead the trajectories of the system state are very different in non-conservative systems; dissipating energy, the system evolves toward a stable equilibrium point; the trajectories cross the energy levels generated in the conservative case (Figure 4.4(right)).

In the previous case we considered as a background hypothesis that the state space in our example is represented on a plane. But a more physical representation exists and is given wrapping the phase portrait onto the surface of a cylinder [9]. This is a "physical" representation of the state space because reproduces the physical space of the moving mass, then it is the natural space state for the pendulum. In addition, as in the physical reality, the two equilibrium points (stable and unstable) are unique.

Consequently on the cylinder, the trajectories of a conservative system are closed curves (Figure 4.5(left)), while the trajectories of a dissipative system are open curves. Starting from initial condition they evolve toward the stable equilibrium point (Figure 4.5(right)).

In a similar way on the cylinder the partitions are simple and direct; the associated automata close naturally on the fundamental states associated to the elements of the partition because on the cylinder the extensions on the left and the right of the plane state space are eliminated. As a consequence the associated automata have the following numbers of states: ($N_F = 8$ and $N_T = 28$) for geometric partition; ($N_F = 3$ and $N_T = 4$) for physical partition. It is evident as we obtained a reduction of the transition states in both the cases of geometric and physical partition, but this reduction is more significative for physical partition (Figure 4.6(left)).

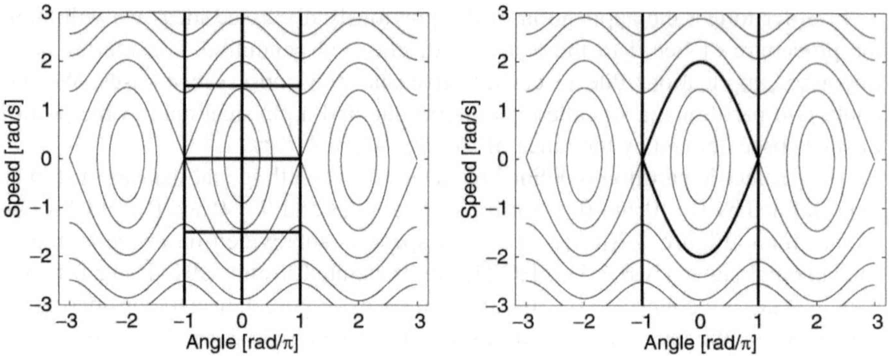

Fig. 4.2: Partition of state space: Geometric partition (left); Physical partition (right)

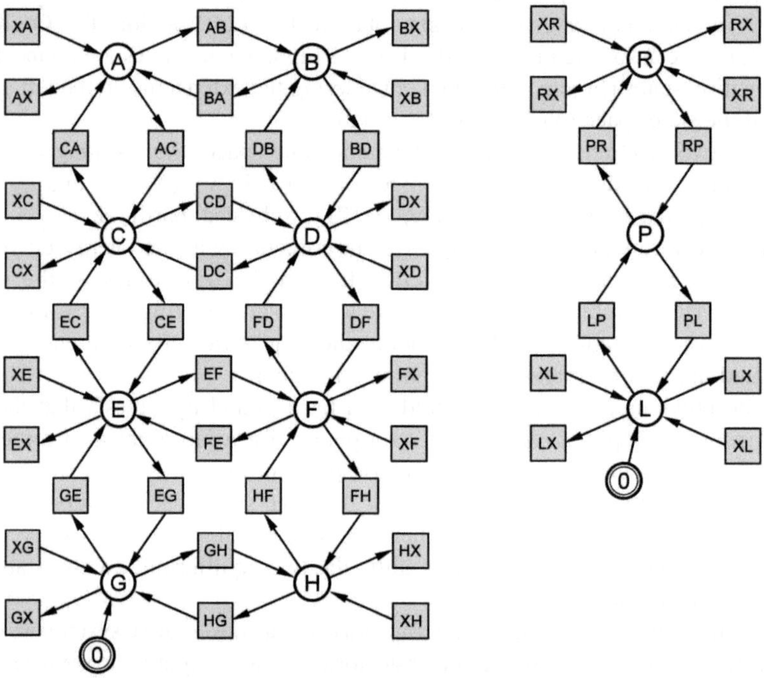

Fig. 4.3: Associated automata: Geometric partition of state space (left); Physical partition of state space (right)

Considering the case of physical partition on the cylinder, it may be convenient to analyze the associated automaton (Figure 4.6(left)) and to give some further definition. Extracting from the associated automaton the transition states and connecting them from an unique initial state, we obtain an automaton collecting only the actions on/from the system (Figure 4.6(right)). Then it may be convenient to introduce the following terms useful to define two descriptive levels:

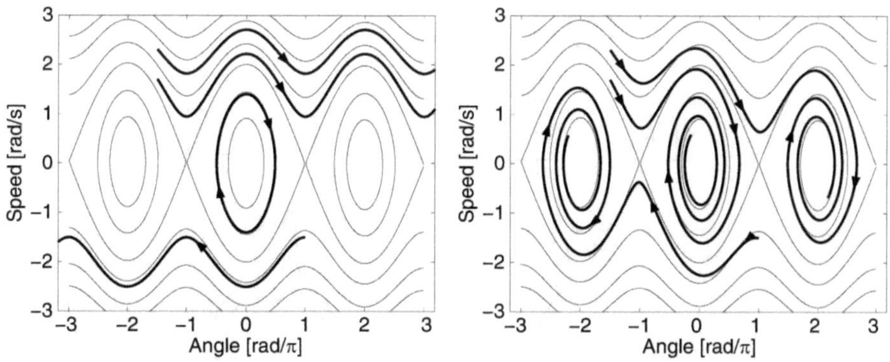

Fig. 4.4: Trajectories in state space: Conservative system (left); Dissipative system (right)

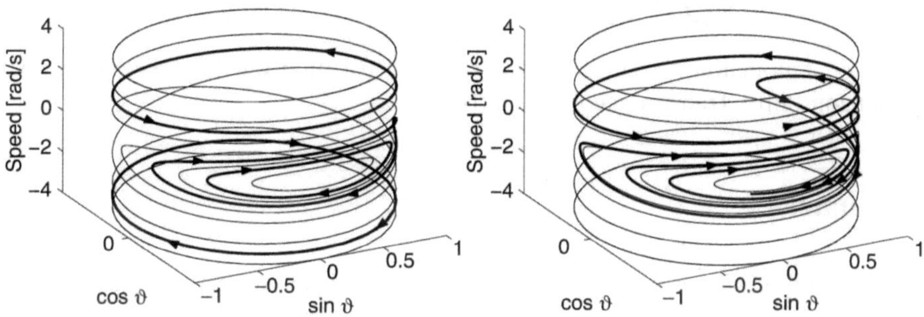

Fig. 4.5: State space on the cylinder: Closed trajectories for a conservative system (left); Open trajectory for a dissipative system (right)

- *Global Automaton*. It is the automaton associated to the dynamical system. It collects all the fundamental and transition states of the system and then it gives a global description of the possible relations between all the states of the system (Figure 4.6(left)).
- *Transitions Automaton*. It collects only the transition states of the system and then it gives a description of the possible actions on the system or resulting from the system (Figure 4.6(right)).

This distinction is useful to represent two different ideas: one is a description of the actions, active and passive (transition states) and the other one is a description of the effects or the results (fundamental states).

In this way we defined two layers in describing associated automata. In a natural way this description can give a connection with the superimposed automata [4], just used in designing and testing of industrial automation and control systems.

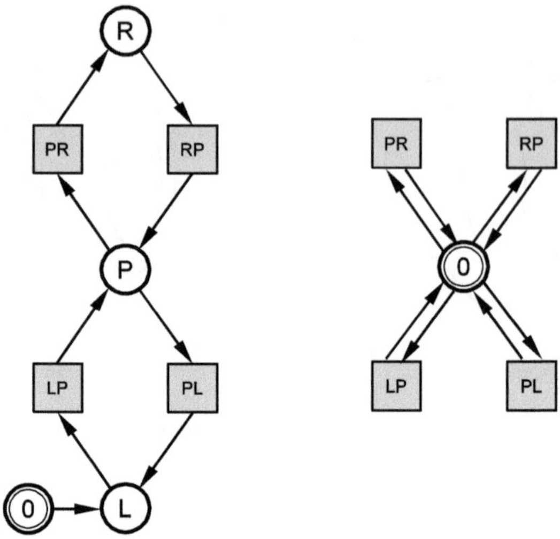

Fig. 4.6: Automaton associated to the physical partition on the cylinder: Global automaton (left); Transition automaton (right)

6 Remarks

Interesting motivations emerge from the example and consequently some considerations are useful to find the correct collocation of this approach.

- Given a dynamical system, classically described by nonlinear differential equations, the method associates an automaton that constitute a discrete and simplified description of the dynamical systems. It can be a natural description of the system, even if we introduce a loss of information as time scales.
- The choice of partition schema is important because it influences the type of information we can force to emerge from the system.

  (1) A mathematical partition supplies a description of the system that puts into evidence the relations between the state variables. The transitions in the system are a composition of the transformations of each state variable.
  (2) A physical partition supplies a description of the system that puts into evidence the relations between the energy levels. The transitions are energy transformations or energy fluxes in the system.

- It is interesting to observe as the transition states in the associated automata to a physical partition correspond exactly with the energy transformations in the system; the interpretation of the transition state, and also of fundamental states, is not so direct and simple for the automata associated to the geometric partition. This difference is more evident if we consider the corresponding representation of the state space on the cylinder. The physics of the system is more evident.

- In the pendulum example, it was natural to use energy level curves in defining the partition of state space. The conservative case constitutes the natural "trama" to build the physical partition of he state space. The non conservative case is depicted in relation to the partition of the conservative case.
  For the nonconservative case other partitions can be defined using the trajectories of the system as border of the cells. The structure of the cells, introducing additional dynamic conditions, can become very complex and is justified only as a description tool.
- The representation of associated automata as superimposed automata may be useful to describe the hierarchical structure of large control systems because the functional architecture of the subsystems can be correctly analyzed, integrated and managed. Then superimposed automata are a natural tool for specifying and designing automation and control systems for many practical applications.

7 Conclusions

- Associated automata are a simplified description of the system that individuates and correlates the fundamental states of the system and shows the possible actions on/from the system. This approach can be very useful in control engineering applications, when the system evolution must be carefully designed and implemented.
- An investigation on the application of this methodology to high order nonlinear systems calls for the development of an adequate theoretical work with the goal to verify and harmonize the available results, methods, examples and design procedures that were developed in many application fields. Among the many investigation directions, it may be interesting to find possible correspondences or relations with hybrid systems formulation [5].
- In designing automation and control systems, engineers need to find a simple high level description for the interpretation of the processes. Following the transformation of energy in the system, intrinsic into state space description, we can concentrate on the key points and the characteristic parameters connected with the natural evolution of the system. Such a condition gives the possibility to build a correct methodology for analyzing, developing, designing and then using "intrinsic safety" in systems.
  These motivations can gain particular interest in designing the Operator Interfaces when it is necessary to relate the multilevel structures for controlling complex processes with the effective synthesis of information necessary to human operator.
- The physical approach, considering the central role of energy, enables to build a "natural" description of the system. This energetic approach is the basis when it is necessary to follow the energy fluxes in a system with the goal of defining effectively its safety conditions. This argument is critical and it becomes a "must" in designing control system based on the utilization of "intrinsic" safety properties of the processes.

References

1. Arnold V.I. (1978). *Ordinary Differential Equations*. The MIT Press, Cambridge, MA.
2. Arnold V.I. (1988). *Geometrical Methods in the Theory of Ordinary Differential Equations*. Springer-Verlag, New York.
3. Arnold V.I. (1989). *Mathematical Methods of Classical Mechanics* (2nd Edition). Springer-Verlag, New York.
4. De Cindio F., De Michelis G., Pomello L., Simone C. and Stragapede A. (1987). *Le Reti di Automi Sovrapposti: Una Classe Modulare di Reti di Petri*. ENEL-CRA, Milano.
5. Goebel R., Sanfelice R.G. and Teel A. (2009). Hybrid Dynamical Systems. *IEEE Control Systems Magazine*, **29**(2): 28–93. DOI: 10.1109/MCS.2008.931718.
6. Flum J., Grädel E. and Wilke T. (Eds.) (2008). *Logic and Automata. History and Perspectives*. Amsterdam University Press, Amsterdam.
7. Jordan D.W. and Smith P. (1999). *Nonlinear Ordinary Differential Equations. An Introduction to Dynamical Systems*, (3rd edition). Oxford University Press, Oxford, UK.
8. Hirsh M.W., Smale S. and Devaney R.L. (2013). *Differential Equations, Dynamical Systems, and an Introduction to Chaos*. Elsevier-Academic Press, Waltham, MA.
9. Strogatz S.H. (1994). *Nonlinear Dynamics and Chaos*. Perseus Books, Cambridge, MA.
10. Susskind L. and Hrabovsky G. (2013). *The Theoretical Minimum. What You Need to Know to Start Doing Physics*. Basic Books, New York.

Part III
The Contribution of Physics to a New General Theory of Systems

Chapter 5
Towards the Study of New Nuclear Energies

Umberto Di Caprio and Mario R. Abram

1 Introduction

In 2012 a contribution dealing with a new process for the generation of new nuclear energies has been published by U. Di Caprio in *Hadronic Journal* [9]. Here we recapitulate the main results. The work deals with the dynamical structure of the deuteron and with the illustration of a chain nuclear reaction.

In Chap. 1 is illustrated the dynamical structure of the deuteron; in Chap. 2 it is explained the meaning of the binding energy; in Chap. 3 the values of the spin and of the electric charge. Finally in Chap. 4 we point out a chain nuclear reaction obtained by disintegrating the deuteron by hitting it with the photon of energy 193 MeV.

2 Dynamical Structure of a Deuteron

We postulate that the deuteron consists of two protons p_1, p_2 (conventionally denominated "external proton" and "antipodic proton") plus an "intermediate" electron, in circular rotation around their center of mass within a plane Π_d (Fig. 5.1 (right)). The rotations are synchronous and, at any time proton p_1 is distant R_1 from the center O, proton p_2 is at the antipodes of p_1, at a distance R_2 from O, with

U. Di Caprio (✉)
ASDE (Associazione Dirigenti Enel), Roma, Italy
e-mail: umbertodicaprio@tiscali.it

M.R. Abram
AIRS (Associazione Italiana per la Ricerca sui Sistemi), Milano, Italy
e-mail: mario.abram@alice.it

G. Minati et al. (eds.), *Towards a Post-Bertalanffy Systemics*, Contemporary Systems Thinking, DOI 10.1007/978-3-319-24391-7_5

$R_2 < R_1$. The electron is in turn at the antipodes of p_1 (and then is aligned with p_2) at a distance R_3 from O such that $R_3 < R_2 < R_1$. Denoting V_1, V_2, V_3 the three rotation speeds we assume

$$V_1 = const\,; \qquad V_2 = const\,; \qquad V_3 = const \tag{5.1}$$

$$\frac{V_3}{R_3} = \frac{V_2}{R_2} = \frac{V_1}{R_1} \ (synchrony). \tag{5.2}$$

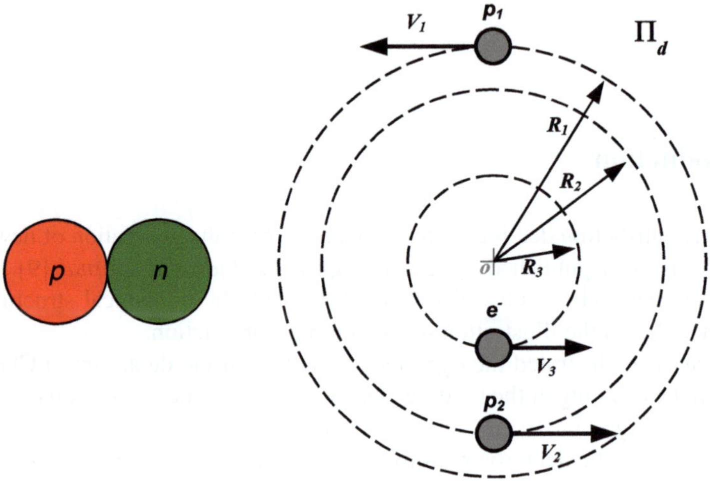

Fig. 5.1: Deuteron structure: SM model of deuteron (*left*), dynamic model of deuteron (*right*)

Another fundamental assumption based upon collateral studies [5, 6], is that each of the two protons is a composite particle formed by three rotating quarks plus a pion, contained within a plane. We assume that planes Π_1 and Π_2 respectively containing p_1 and p_2 form an angle $\psi = 30°$ with plane Π_d of the deuteron (i.e. the horizontal plane of motion of the "point-form" particles p_1, p_2, e). We also assume that in the three-dimensional structure of the deuteron the electron remains pointform (within an approximation of 10^{-19} m) and that its spin and its magnetic moment are perpendicular to Π_d. At any time the center of mass of composite particle p_1 belongs to the intersection of Π_1 with Π_d and represents a mobil point on the circle with radius R_1 and center O (in plane Π_d). Analogously, the center of mass of composite particle p_2 belongs to intersection of Π_2 with Π_d and represents a mobil point on a circle with radius R_2 and center O (in plane Π_d). All in all, the deuteron has a *three-dimensional structure* (Fig. 5.2) and in plane Π_d the distance between the centers of mass of the two protons is in the order 10^{-18} m. It is clear that no overlapping among particles exists in spite of the fact the proton radius (which is about $1.4 \cdot 10^{-15}$ m) is larger than the *distance* between Π_1 and Π_2.

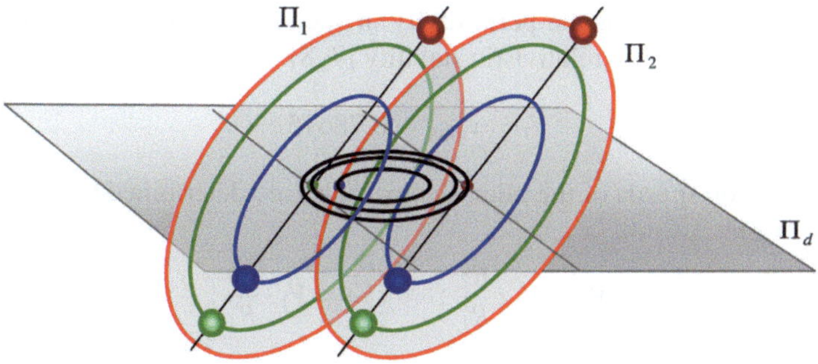

Fig. 5.2: Three dimensional structure of the deuteron

We call *solution of the deuteron problem* any set of values R_1, R_2, R_3, V_1, V_2, V_3, ψ so that the three-body structure in object gives the experimental values of charge, mass, magnetic moment, spin of the deuteron. In addition we want to explain the character of the strong interaction and the experimental value $g_d \sim 14$. Furthermore, we require that the three-body dynamic structure in object be *stable.*

To solve the problem, we use relativistic equations derived from Special Relativity (SR) and the principle of equivalence Potential energy/mass. In particular, for each particle we determine its contribution to the total mass by equations of type

$$\hat{m} = \gamma\left(m + \frac{E_p}{c^2}\right) \tag{5.3}$$

where γ is the relativistic coefficient

$$\gamma = \frac{1}{\sqrt{1-(v^2/c^2)}} \tag{5.4}$$

and E_p is the Coulomb potential energy. So we have three distinct contributions, one for each particle and, in vie of determining total mass we add a fourth contribution (E_H/c^2) that accounts for the inertial effects of the binding electromagnetic energy. The meaning of E_H is explained in Sect. 3. All in all, the mass of the deuteron is given by

$$m_d = (m_1 + m_2 + m_3) + m_4 \tag{5.5}$$

$$m_1 = \gamma_1\left[m_p + \frac{E_{12}}{c^2} + \frac{E_{13}}{c^2}\right]; \quad m_2 = \gamma_2\left[m_p + \frac{E_{12}}{c^2} + \frac{E_{23}}{c^2}\right] \tag{5.6}$$

$$m_3 = \gamma_3\left[m_0 + \frac{E_{13}}{c^2} + \frac{E_{23}}{c^2}\right]; \quad m_4 = \frac{E_H}{c^2}$$

$$E_{12} = \frac{kq^2}{R_1+R_2}; \qquad E_{13} = -\frac{kq^2}{R_1+R_3}; \qquad E_{23} = -\frac{kq^2}{R_2-R_3} \tag{5.7}$$

where $k = 1/(4\pi\varepsilon_0)$; ε_0 is the permittivity in vacuum. The rotation speed of p_1 is the maximum allowed by relativistic stability [7, 8]

$$V_1 = v_{\max} = 0.78615\,c \tag{5.8}$$

while the rotation speeds of the other two particles are determined by the isochrony condition (5.2):

$$V_2 = V_1 \cdot \frac{R_2}{R_1}\,; \qquad V_3 = V_1 \cdot \frac{R_3}{R_1}\,. \tag{5.9}$$

The relativistic coefficients γ_1, γ_2, γ_3 are given by

$$\gamma_1 = \frac{1}{\sqrt{1 - V_1^2/c^2}} = \gamma_{\max} = 1.618 \tag{5.10}$$

$$\gamma_2 = \frac{1}{\sqrt{1 - V_2^2/c^2}}\,; \qquad \gamma_3 = \frac{1}{\sqrt{1 - V_3^2/c^2}}\,.$$

We claim that the following set of values identifies a *solution of the deuteron problem*

$$R_1 = 8.955 \cdot 10^{-18}\,\text{m}; \quad R_2 = 4.835 \cdot 10^{-18}\,\text{m}; \quad R_3 = 0.915 \cdot 10^{-18}\,\text{m} \tag{5.11}$$

$$V_1 = 0.78615 \cdot c; \qquad V_2 = 0.42446 \cdot c; \qquad V_3 = 0.0803 \cdot c \tag{5.12}$$

$$\psi = 0.786\,[\text{rad}]. \tag{5.13}$$

In fact, they give us

$$E_{12} = 104.41\,[\text{MeV}]\,; \qquad E_{13} = -145.9\,[\text{MeV}]\,; \qquad E_{23} = -367.3\,[\text{MeV}] \tag{5.14}$$

$$\gamma_1 = 1.618; \qquad \gamma_2 = 1.1044; \qquad \gamma_3 = 1.0032. \tag{5.15}$$

Moreover, (see Sect. 3) the electromagnetic energy E_H turns out equal to

$$E_H = 193\,[\text{MeV}]. \tag{5.16}$$

Then

$$m_1 = \gamma_1\,(938.27 + 104.46 - 145.9)\ [\text{MeV}/c^2] = \gamma_1\ 896.83\ [\text{MeV}/c^2] \tag{5.17}$$

$$m_2 = \gamma_2\,(938.27 + 104.46 - 367.33)\ [\text{MeV}/c^2] = \gamma_2\ 675.39\ [\text{MeV}/c^2] \tag{5.18}$$

$$m_3 = \gamma_3\,(0.511 - 145.9 - 367.33)\,[\text{MeV}/c^2] = \gamma_3\ (-512.72)\ [\text{MeV}/c^2] \tag{5.19}$$

$$\begin{aligned} m_1 + m_2 + m_3 &= 1.618 \times 896.83 + 1.1044 \times 675.39 + 1.0032 \times (-512.72) \\ &= 1682.57\ [\text{MeV}/c^2] \end{aligned} \tag{5.20}$$

$$(m_1 + m_2 + m_3) + m_4 = (1682.57) + E_H/c^2 = 1875.57\ [\text{MeV}/c^2] \equiv m_d\,.$$

This proves that the total relativistic mass is equal to the experimental value. Also,

$$m_1 + m_3 = 937.3\ [\text{MeV}] = m_n - 2.26\ [\text{MeV}/c^2] \tag{5.21}$$

$$m_2 + E_H/c^2 = 938.27\ [\text{MeV}/c^2] = m_p \tag{5.22}$$

with $m_n = 939.56\ [\text{MeV}/c^2]$ mass of the neutron. That fully explains the so-called mass deficit of the deuteron.

With regards to the magnetic moment of the deuteron, which is a vector quantity, we assume that the direction is perpendicular to planes Π_1 and Π_2, and since these planes form an angle ψ with the deuteron plane Π_d, we further assume that the magnitude of $\overline{\mu}_d$ is given by

$$\mu_d = \mu_p + \mu_n + \left(-\frac{qV_1R_1}{2} + \frac{qV_2R_2}{2} + \frac{qV_3R_3}{2} \sin\psi \right) \tag{5.23}$$

where

$$\psi = \frac{\pi}{6} = 0.5236 = 30^\circ ; \qquad \sin\psi = 0.5 \tag{5.24}$$

$$\mu_p = 2.7928 \frac{q\hbar}{2m_p} ; \qquad \mu_n = -1.913 \frac{q\hbar}{2m_p} . \tag{5.25}$$

As

$$\frac{qV_1R_1}{2} = 0.0334 \frac{q\hbar}{2m_p} ; \ \frac{qV_2R_2}{2} = 0.00979 \frac{q\hbar}{2m_p} ; \ \frac{qV_3R_3}{2} = 3.45 \times 10^{-4} \frac{q\hbar}{2m_p} \tag{5.26}$$

we find

$$\mu_d = \frac{(2.7928 - 1.913) - 0.0236 + 3.45 \cdot 10^{-4} \sin\psi}{2m_p} q\hbar = 0.85637 \frac{q\hbar}{2m_p} \tag{5.27}$$

in accordance with the experimental value.

The quantity within brackets in Eq. (5.24) represents the dynamical contribution to the deuteron's magnetic moment owing to the rotation of the three point-form particles that form the deuteron in plane Π_d. The direction of $\overline{\mu}_d$ is the famous z direction of particle physics.

We shall subsequently prove that the spin of the deuteron in the direction z is $\hbar$, provided that the direction of the electron spin, as well as that of the electron's magnetic moment, is z (Fig. 5.4).

3 The Meaning of the Electromagnetic Energy (Binding Energy)

Any proton is formed by three rotating particles (with charges $2q/3$, $-q/3$, $2q/3$) and hence generates a magnetic field (Fig. 5.3(right)) [4, 5]. (The three fields are responsible for the well-known magnetic moment of the proton and for quantization of the electron orbits in hydrogen.) In particular, the two protons of the deuteron give rise to two magnetic fields H_1 and H_2 which, owing to the structure of the model are "rotating" fields. So, they induce an electromotive force on the electron and the related electromagnetic energy is:

$$E_H = E_{H_1} + E_{H_2} \tag{5.28}$$

$$E_{H_1} = (\mu_0 H_1) \cdot \frac{qV_1 R_3}{2}\,; \qquad E_{H_2} = (\mu_0 H_2) \cdot \frac{qV_2 R_3}{2} \tag{5.29}$$

μ_0 magnetic permeability in vacuum, with

$$H_1 = \frac{[(2q/3)\,v_1 r_1 - (q/3)\,v_2 r_2 + (2q/3)\,v_3 r_3]}{4\pi(R_1 - R_3)^3} \cdot f(\psi) \tag{5.30}$$

$$H_2 = \frac{[(2q/3)\,v_1 r_1 - (q/3)\,v_2 r_2 + (2q/3)\,v_3 r_3]}{4\pi(R_2 - R_3)^3} \cdot f(\psi) \tag{5.31}$$

where, for each proton, r_1, r_2 and r_3 are the distances of the three constituent particles from the center of mass of the proton while v_1, v_2 and v_3 are the rotation speeds of such particles (see Fig. 5.1(right)). Also ψ is the angle formed by each of the two protons planes with the plane of the deuteron Π_d and according to preceding assumptions $\psi = 30°$ (Fig. 5.2).

The numerical values of the aforesaid quantities are reported in a study by Di Caprio about the dynamical structure of the proton [5]

$$r_1 = 1.399 \times 10^{-15}\,[\mathrm{m}]; \quad r_2 = 1.3788 \times 10^{-15}\,[\mathrm{m}]; \quad r_3 = 1.4 \times 10^{-15}\,[\mathrm{m}] \tag{5.32}$$

$$v_1 = 0.95775\,c; \qquad v_2 = 0.94477\,c; \qquad v_3 = 0.9599\,c. \tag{5.33}$$

The form of function $f(\psi)$ can be deducted from the numerical analysis of the general results on electro-magnetism illustrated by Arenhövel in [1, 2]

$$f(\psi) = \left[1 + \tan\left(\frac{\pi}{2} - \psi\right)\right] \cdot \left[\tan\left(\frac{\psi}{2}\right)\right]^{32/9}. \tag{5.34}$$

Inserting $\psi = \pi/6$ we find the numerical value $f(\psi) = 0.025$. With the above values (and with R_1, R_2, R_3, V_1, V_2, V_3 as given by (5.11), (5.12))

$$E_{H_1} = 5.4664 \times 10^{-12}\,[\mathrm{J}] = 34.12\,[\mathrm{MeV}] \tag{5.35}$$

$$E_{H_2} = 25.4 \times 10^{-12}\,[\mathrm{J}] = 158.93\,[\mathrm{MeV}] \tag{5.36}$$

$$E_H = E_{H_1} + E_{H_2} = 193\,[\mathrm{MeV}] \tag{5.37}$$

in accordance with (5.16).
We claim that E_H is but the binding energy E_B, namely

$$E_B = E_H = 193\,[\mathrm{MeV}]. \tag{5.38}$$

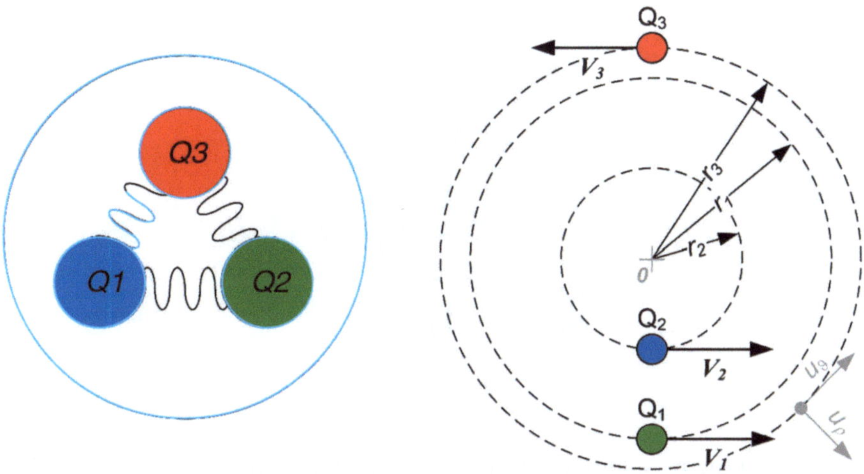

Fig. 5.3: Proton structure: SM model of the proton (*left*), dynamic model of the proton (*right*)

As a matter of fact, setting

$$g_d = \frac{1}{\alpha} \frac{E_B}{m_d c^2} \tag{5.39}$$

where α is the *fine structure constant* we find

$$g_d = 137 \cdot \frac{193\,[\mathrm{MeV}]}{1875.57\,[\mathrm{MeV}]} = 14.1 \sim 14\,. \tag{5.40}$$

So, g_d is the experimental value of the *strong interaction constant*. In second place, 193 MeV is the experimental value of a resonance energy for photodisintegration of the deuteron [3].

4 The Spin and the Electric Charge

The deuteron spin is $s_d = \hbar$. In our model the spin is the sum of three contributions owing to the spins of three particles (two protons and an electron). Furthermore, the spin is a vectorial quantity. Considering the direction z identified by the perpendicular to the plane of motion of the three pointform particles we find that the spin of the deuteron in the direction z is given by (see Fig. 5.4)

$$spin\, d\,(in\, the\, direction\, z) = spin\, p_1 \cdot \sin\psi + spin\, p_2 \cdot \sin\psi + electron\, spin \tag{5.41}$$

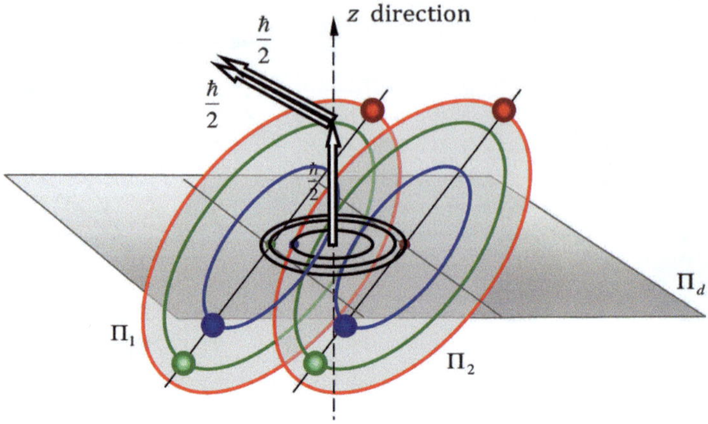

Fig. 5.4: Deuteron spin: vectorial composition of the spins of the three particles

with $\psi = \pi/6$. Therefore,

$$spin\,d\,(in\,the\,direction\,z) = \frac{\hbar}{2}\sin\frac{\pi}{6} + \frac{\hbar}{2}\sin\frac{\pi}{6} + \frac{\hbar}{2} = \frac{\hbar}{4} + \frac{\hbar}{4} + \frac{\hbar}{2} = \hbar . \qquad (5.42)$$

This sum equals $\hbar$ in agreement with the experimental value.
As regards the electric charge we have

$$charge\ of\ deuteron = q + q - q = q \qquad (5.43)$$

which is o.k.

5 Chain Nuclear Reaction

The energized proton ($\gamma_1 p_1$) has a kinetic energy about three times the binding energy of the deuteron. In fact,

$$T_1 = (\gamma_1 - 1)\, m_p \cdot c^2 = 0.618 \cdot 938.27\ [\mathrm{MeV}] = 579.97\ [\mathrm{MeV}] \qquad (5.44)$$

while

$$3 \cdot E_B = 3 \cdot 193\ [\mathrm{MeV}] = 579\ [\mathrm{MeV}]. \qquad (5.45)$$

Consequently,

$$T_1 \sim 3 \cdot E_B\,. \qquad (5.46)$$

That means that after disintegrating a deuteron by hitting it with a photon of energy 193 MeV we get an energized proton (plus a particle η and a particle ν_e) which in turn can break three more deuterons, and so on. The reaction is divergent and then, in this manner, we can produce clean nuclear energy, provided we control the

reaction. The control can be eventually achieved with the use of cadmium bars the purpose of which is simply to intercept part of the energized proton (not to moderate neutrons, since this is not the philosophy of our fission procedure; it is clear that we are dealing with fission, not with fusion). The following example illustrates a controlled chain reaction in two stages.
STAGE 1: fission of a first nucleus according to

$$193\,[\mathrm{MeV}] + d_1 = 2068.6\,[\mathrm{MeV}] \rightarrow 1.618\,m_p + \eta + 3e^+ + 3e^- \tag{5.47}$$

$$\eta = 547.5\,[\mathrm{MeV}] \tag{5.48}$$

$$1.618\,m_p + \eta \rightarrow p + \eta + 0.618\,m_p = p + \eta + 578.86\,[\mathrm{MeV}]. \tag{5.49}$$

STAGE 2: fission of a second nucleus, according to

$$578.86\ [\mathrm{MeV}] + d_2 = 2455.45\ [\mathrm{MeV}] \rightarrow \Omega^- + 2e^+ + 193\ [\mathrm{MeV}] \tag{5.50}$$

$$\Omega^- = 2261\ [\mathrm{MeV}]. \tag{5.51}$$

SUBSEQUENT Decay

$$\Omega^- = N_{1530} + K^- + 236\ [\mathrm{MeV}] \tag{5.52}$$

Released energy 236 [MeV].

6 The Center of Mass

The following equation proves that the geometric center of the deuteron (point O) is the center of mass:

$$m_1\,(r_2 - r_3) = m_{23}\cdot r_1 + \frac{E_H}{c^2}\,(r_2 - r_3\cos\psi) \tag{5.53}$$

with

$$m_{23} = \frac{m_2 r_2 + m_3 r_3}{r_2 + r_3}. \tag{5.54}$$

We can check that Eq. (5.54) turns out satisfied. In fact it is

$$m_1 = 1.618\cdot 896.836\ [\mathrm{MeV}/c^2] = 1451.07\ [\mathrm{MeV}/c^2] \tag{5.55}$$

$$m_2 = 1.1044\cdot 675.39\ [\mathrm{MeV}/c^2] = 745.9\ [\mathrm{MeV}/c^2] \tag{5.56}$$

$$m_3 = 1.0032\cdot(-512.79)\ [\mathrm{MeV}/c^2] = -514.38\ [\mathrm{MeV}/c^2] \tag{5.57}$$

and then

$$m_{23} = \frac{745.9 \cdot 4.835 - 514.38 \cdot 0.915}{4.835 + 0.915} \,[\mathrm{MeV}/c^2] \tag{5.58}$$
$$= 545.35\,[\mathrm{MeV}/c^2]$$
$$m_1\,(r_2 - r_3) = 1451.07 \cdot 3.92 \times 10^{-18}\,[\mathrm{MeV}/c^2] \cdot m \tag{5.59}$$
$$= 5688 \times 10^{-18} [\mathrm{MeV}/c^2] \times m$$
$$m_{23} \cdot r_1 = 545.35 \cdot 8.955 \times 10^{-18}\,[\mathrm{MeV}/c^2] \cdot m \tag{5.60}$$
$$= 4883.6 \times 10^{-18}\,[\mathrm{MeV}/c^2] \cdot m$$
$$\frac{E_H}{c^2}\,(r_2 - r_3 \cos\psi) = 193\,[\mathrm{MeV}/c^2] \cdot 4.17 \times 10^{-18} m \tag{5.61}$$
$$= 805.47 \times 10^{-18}\,[\mathrm{MeV}/c^2] \cdot m.$$

It follows from the above equations that

$$m_{23} \cdot r_1 + \frac{E_H}{c^2}\,(r_2 - r_3 \cos\psi) = (4883.6 + 805.47) \times 10^{-18}\,[\mathrm{MeV}/c^2] \cdot m$$
$$= 5688 \times 10^{-18}\,[\mathrm{MeV}/c^2] \cdot m. \tag{5.62}$$

Equations (5.60) and (5.63) bring about that Eq. (5.54) is satisfied.

7 Remark

Our process is strictly based on *fission* and has *nothing to do with fusion*. By virtue of the relatively low temperatures involved, undesirable and accidental fusions can be excluded. The existing technology for generating electric energy using conventional thermal or even nuclear power plants can easily be utilized. Of course, there is no question about the necessity of convenient and conclusive experiments regarding the practical start of the chain reaction.

8 Conclusions

Our theoretical analysis joined with experimental results (in particular see [3]) prove that the deuteron is formed by three rotating particles, two protons and one electron. The rotations are synchronous and their geometric center coincides with the center of mass. The basic properties of the deuteron are satisfactory reproduced by a three-body relativistic model that marks a substantial advancement with respect to the static two-body standard model (SM) utilized by the international scientific community. We have clearly explained the reasons why the SM fails to indicate correct values for the mass and magnetic moment, and assumes, erroneously, that the binding energy equals the so called mass deficit (multiplied c^2) which, unfortunately, is negative. We have also seen the inefficacy of the SM in view of explaining the various manners of decomposition of the deuteron by the action of convenient excitation energies. In addition, it does not clarify the nature of the tie between the proton and the neutron nor does it prove the stability of such a tie. On the contrary, we have been able to show the stability of the three-particle model and the fact that the binding energy is *positive* and equal to 193 MeV (against −2.26 MeV). Hitting the deuteron with a photon, carrying this amount of energy, we disintegrate the deuteron and eventually trigger a nuclear chain reaction with a gain factor equal to three. We have also identified all of the other ways of decomposition observed in experiments, so that one now has a unique theoretical frame available to explain MEC (meson exchange) and bearing of particles ρ, Δ, N, η, η'. The core of the study is, however, the demonstration that the nuclear force is an electromagnetic force, counteracted by inertial force. Moreover, the binding energy is the electromagnetic energy acquired by the electron because of its rotation in the magnetic field generated by the two protons. By using these fundamental new notions and, in parallel, adopting relativistic equations, we can easily extend the study to other light nuclei, e.g. carbon.

Appendix: Stability Condition

The equation

$$\frac{\gamma v^2}{c^2} = 1 \tag{5.63}$$

determines the rotation speed and the relativistic mass coefficient of proton p_1 in fact as

$$\gamma = \frac{1}{\sqrt{1 - v^2/c^2}} \tag{5.64}$$

it is

$$\frac{v^2}{c^2} = \frac{\gamma^2 - 1}{\gamma^2} \qquad \frac{\gamma v^2}{c^2} = \gamma^2 - \gamma \tag{5.65}$$

so that Eq. (5.64) results in

$$\gamma^2 = \gamma + 1 \tag{5.66}$$

which is the classical equation of the golden ratio, namely the positive solution is

$$\gamma = \gamma_r = \frac{1 + \sqrt{5}}{2} = 1.618. \tag{5.67}$$

So

$$\frac{\gamma_r v_r^2}{c^2} = 1 \tag{5.68}$$

and v_r, γ_r identify the rotation speed and the relativistic mass coefficient of particle p_1, in accordance with our general assumptions. Now, we show that condition (5.64) is sufficient for the stability of the motion of p_1 and consequently for the stability of the remaining two particle p_2 and e, which rotate synchronously with p_1.

References

1. Arenhövel, H. (1992). Subnuclear degrees of freedom in electromagnetic interactions in nuclei. *Chinese Journal of Physics, 30*, 17–95.
2. Arenhövel, H., & Drechsel, D. (Eds.). (1979). *Nuclear Physics with Electromagnetic Interactions. Proceedings of the International Conference*, Held in Mainz, Germany, 5–9 June 1979. Berlin: Springer.
3. Arenhövel, H., & Sanzone, M. (2003). *Photodisintegration of the deuteron. A review of theory and experiment*. Wien/New York: Springer (f.ed. 1991, reed. 2003).
4. Di Caprio, U. (2000). The effects of the proton's magnetic field upon quantization. *Hadronic Journal, 23*, 689–704.
5. Di Caprio, U. (2001). The dynamic structure of the proton. *Hadronic Journal Supplement,16*, 163–182.
6. Di Caprio, U. (2001). *The dynamic structure of the neutron*. Int. Rep. SA-2, Stability Analysis, Milano.
7. Di Caprio, U. (2009). Relativistic stability. Part 1 – Relation between special relativity and stability theory in the two-body problem. In G. Minati, M. Abram, & E. Pessa (Eds.), *Processes of emergence of systems and systemic properties* (pp. 659–672). Singapore: World Scientific.
8. Di Caprio, U. (2009). Relativistic stability. Part 2 – A study of black holes and of the Schwarzschild radius. In G. Minati, M. Abram, & E. Pessa (Eds.), *Processes of emergence of systems and systemic properties* (pp. 673–684). Singapore: World Scientific.
9. Di Caprio, U. (2012). Study of the electromagnetic content of Santilli's three body structure model of the Deuteron and its implications for new nuclear energies. *Hadronic Journal, 35*(2), 125–174.

Chapter 6
Decomposing Dynamical Systems

Marco Giunti

1 Introduction

A dynamical system is a kind of mathematical model that is intended to capture the intuitive notion of an arbitrary deterministic system, either reversible or irreversible, with discrete or continuous time or state space [1, 2, 4, 5]. Giunti and Mazzola [3] generalize the standard notion of a dynamical system by simply taking the state space M to be a non-empty set, and by requiring for the time set T a quite simple algebraic structure—namely, a monoid $L = (T, +)$, which they call the time model. They thus define a dynamical system DS_L on a monoid L, which they claim to be the minimal mathematical model that still captures the intuitive idea of a deterministic dynamics; such a model consists of a state space M together with a family, indexed by T, of functions from M to M, which satisfy an identity and a composition condition.

Cellular automata are well known examples of dynamical systems with discrete time set (the non-negative integers Z^+) and discrete state space. Wolfram [6] showed that the state space of a particular finite cellular automaton (*rule* 90_{10}, see Fig. 6.1) can be exhaustively decomposed into seventy mutually disconnected constituents. Each constituent subspace is internally connected, and it turns out to be the state space of a subsystem of the cellular automaton; the complete dynamics of the cellular automaton can thus be obtained as the sum of the dynamics of its seventy constituent systems (see Fig. 6.2).

Wolfram's example is not peculiar to the quite special case he analyzed. In fact, I will show in this paper that the state space of any dynamical system DS_L on a monoid L can be exhaustively decomposed into a set of mutually disconnected constituents (Decomposition Theorem, Theorem 1), where each constituent is internally

M. Giunti (✉)
ALOPHIS, Applied Logic Philosophy and History of Science,
University of Cagliari, Via Is Mirrionis 1, 09123 Cagliari, Italy
e-mail: giunti@unica.it

G. Minati et al. (eds.), *Towards a Post-Bertalanffy Systemics*, Contemporary Systems Thinking, DOI 10.1007/978-3-319-24391-7_6

Twelve cells arranged in a circle. Each cell may take value 0 or 1. Thus, this CA has 2^{12} = 4096 possible states.

Rule	111	110	101	100	011	010	001	000
	0	1	0	1	1	0	1	0

Rule number (base 2) $01011010_2 = 90_{10}$ (base 10)

Time 0	0	1	0	0	1	0	1	0	1	1	1	1
Time 1	0	0	1	1	0	0	0	0	1	0	0	1

Fig. 6.1: A finite cellular automaton with 12 cells—*rule* 90_{10}

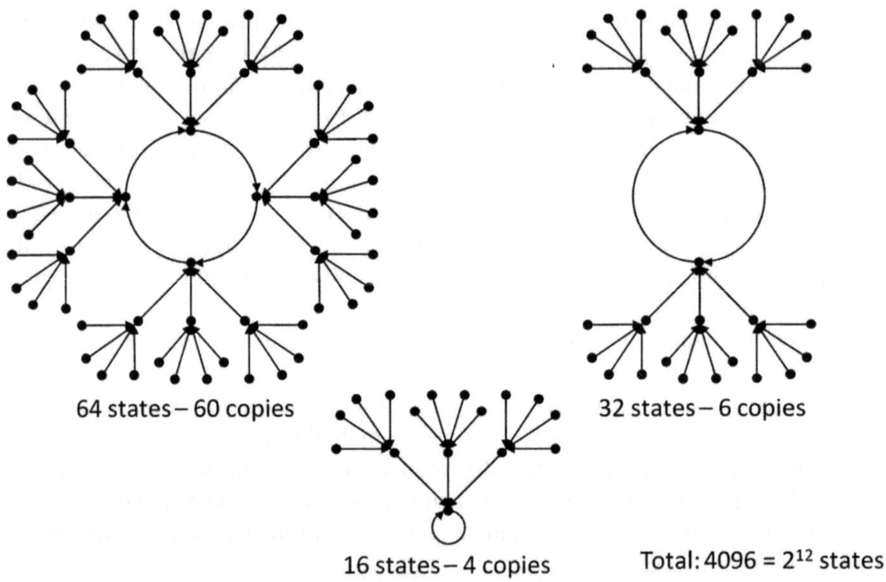

Fig. 6.2: The seventy constituent systems of finite CA *rule* 90_{10}

connected and is the state space of a subsystem of DS_L (called a constituent system of DS_L). In addition, constituent systems are themselves indecomposable (Proposition 6), even though they may very well be complex. Finally, I will show that any dynamical system DS_L is in fact identical to the sum of all its constituent systems (Composition Theorem, Theorem 2). Constituent systems can thus be thought as the indecomposable, but possibly complex, building blocks to which the dynamics of an arbitrary complex system fully reduces. However, no further reduction of the constituents is possible, even if they are themselves complex.

2 Dynamical Systems on Monoids

As mentioned above, a dynamical system is a kind of mathematical model that purports to formally capture the intuitive notion of an arbitrary deterministic system, either reversible or irreversible, with discrete or continuous time or state space. Let $T = Z^+$ (non-negative integers), Z (integers), R^+ (non-negative reals), or R (reals). Below is a standard definition of a dynamical system [1, 2, 4, 5].

Definition 1 (DYNAMICAL SYSTEM). *DS is a dynamical system* := *DS* is a pair $(M, (g^t)_{t \in T})$ such that

1. T is either Z^+, Z, R^+, or R;
   - any $t \in T$ is called *a duration*, and T *the time set* of *DS*;
2. M is a non-empty set;
   - any $x \in M$ is called *a state*, and M *the state space* of *DS*;
3. $(g^t)_{t \in T}$ is a family, indexed by T, of functions from M to M;
   - for any $t \in T$, g^t is called *the (state) transition of duration t* or, briefly, *the t-transition* or *the t-advance* of *DS*;
4. for any $v, t \in T$, for any $x \in M$,
   a. $g^0(x) = x$;
   b. $g^{v+t}(x) = g^v(g^t(x))$.

Example 1. The following are all examples of dynamical systems.

1. Discrete time set $(T = Z^+)$ and discrete state space: Finite state machines, Turing machines, cellular automata restricted to finite configurations.[1]
2. Discrete time set $(T = Z^+)$ and continuous state space: Many systems specified by difference equations, iterated mappings on R, unrestricted cellular automata.
3. Continuous time set $(T = R)$ and continuous state space: Systems specified by ordinary differential equations, many neural networks.

Giunti and Mazzola [3] point out that the standard definition of a dynamical system (Definition 1) is not fully explicit, for it does not make clear exactly which

[1] The state space of a cellular automaton is discrete (i.e. finite or countably infinite) if the cellular automaton state space only includes finite configurations, that is to say, configurations where all but a finite number of cells are in the quiescent state. If this condition is not satisfied, the state space has the power of the continuum.

structure on the time set T is needed in order to support appropriate dynamics. By condition 1 of Definition 1, T is either Z^+, Z, R^+, or R. With respect to the addition operation, these four models share the structure of a linearly ordered commutative monoid; but it is by no means obvious that all this structure on T is needed for a general definition of a dynamical system.

Giunti and Mazzola [3] maintain that the minimal structure on the time set that makes for a materially adequate definition of a dynamical system is just that of a *monoid*. Accordingly, they generalize Definition 1 as follows.

Definition 2 (DYNAMICAL SYSTEM ON A MONOID). *DS_L is a dynamical system on L* $:=$ DS_L is a pair $(M, (g^t)_{t\in T})$ and L is a pair $(T, +)$ such that

1. L is a monoid;
   - any $t \in T$ is called *a duration*, T *the time set*, and L *the time model* of DS;
2. M is a non-empty set;
   - any $x \in M$ is called *a state*, and M *the state space* of DS_L;
3. $(g^t)_{t\in T}$ is a family, indexed by T, of functions from M to M;
   - for any $t \in T$, g^t is called *the (state) transition of duration t* or, briefly, *the t-transition* or *the t-advance* of DS_L;
4. for any $v, t \in T$, for any $x \in M$,

   a. $g^0(x) = x$, where 0 is the unity of L;
   b. $g^{v+t}(x) = g^v(g^t(x))$.

It is interesting to realize that any vector space VS_F over a field F (and, even more generally, any unital left module US_R over a ring with unity R) turns out to be a dynamical system DS_L on a monoid L, where appropriate further structure has been added to both DS_L and L. Thus, the theory of dynamical systems on monoids is a natural generalization of (1) the theory of vector spaces over fields, and (2) the theory of unital left modules over rings with unity.[2]

3 Subspaces and Subsystems

If $DS_L = (M, (g^t)_{t\in T})$ is a dynamical system on a monoid $L = (T, +)$, a subspace of the state space M is any non-empty subset of M which is closed under all t-transitions g^t. More formally,

Definition 3. Let $DS_L = (M, (g^t)_{t\in T})$ be a dynamical system on a monoid $L = (T, +)$.

X is a subspace of M $:= X \subseteq M, X \neq \emptyset$ and, $\forall x \in X, \forall t \in T, g^t(x) \in X$.

[2] My thanks to Tomasz Kowalski for pointing out to me the relation between vector spaces over fields and dynamical systems on monoids.

Proposition 1 *Let $DS_L = (M, (g^t)_{t \in T})$ be a dynamical system on a monoid $L = (T, +)$; let $X \subseteq M$, and g^t/X be the restriction of g^t to X.*

$(X, (g^t/X)_{t \in T})$ is a dynamical system on L iff X is a subspace of M.

Proof. Suppose that $(X, (g^t/X)_{t \in T})$ is a dynamical system on $L = (T, +)$; then, by 2 of Definition 2, $X \neq \emptyset$ and, by 3 of Definition 2, $\forall x \in X, \forall t \in T, g^t/X(x) \in X$. As g^t/X is the restriction of g^t to X, $\forall x \in X, \forall t \in T, g^t(x) = g^t/X(x) \in X$. Hence, as $X \subseteq M$, by Definition 3, X is a subspace of M.

Conversely, suppose that X is a subspace of M. Then, by Definition 3, condition 2 of Definition 2 is satisfied. As g^t/X is the restriction of g^t to X, and by Definition 3, condition 3 of Definition 2 is satisfied as well. Finally, as DS_L is a dynamical system on L and g^t/X is the restriction of g^t to X, conditions 4a and 4b of Definition 2 also hold. Therefore, $(X, (g^t/X)_{t \in T})$ is a dynamical system on $L = (T, +)$. □

A subsystem of a dynamical system $DS1_L = (M, (g^t)_{t \in T})$ on a monoid $L = (T, +)$ is a dynamical system $DS2_L = (N, (h^t)_{t \in T})$ on the same monoid L, whose state space N is a subset of M and whose t-transitions are the restrictions to N of the t-transitions of $DS1_L$. That is to say,

Definition 4. Let $DS1_L = (M, (g^t)_{t \in T})$ be a dynamical system on a monoid $L = (T, +)$.

$DS2_L$ is a subsystem of $DS1_L$:= $DS2_L = (N, (h^t)_{t \in T})$ is a dynamical system on L, $N \subseteq M$ and, $\forall t \in T$, $h^t = g^t/N$.

Obviously, by Definition 4, any dynamical system DS_L on a monoid L is a subsystem of itself. Furthermore, the following proposition holds.

Proposition 2 *Let $DS1_L = (M, (g^t)_{t \in T})$ be a dynamical system on a monoid $L = (T, +)$.*

$DS2_L$ is a subsystem of $DS1_L$ iff $DS2_L = (N, (h^t)_{t \in T})$, N is a subspace of M, and $\forall t \in T$, $h^t = g^t/N$.

Proof. The thesis easily follows from Definitions 4, 3 and Proposition 1. □

The concept of a subsystem allows us to introduce a quite general notion of a simple system. In this acceptation, a dynamical system is simple if it does not possess any proper subsystem; otherwise, it is complex.

Definition 5. Let $DS_L = (M, (g^t)_{t \in T})$ be a dynamical system on a monoid $L = (T, +)$.

DS_L is simple := DS_L does not have any subsystem but itself.

Example 2. Let $DS_L = (\{x\}, (id^t)_{t \in Z^+})$, where x is an arbitrary object and, $\forall t \in Z^+$, id^t is the t-th iteration of the identity function on $\{x\}$. Then, DS_L is a dynamical system on $L = (Z^+, +)$ and DS_L is simple.

Definition 6. Let $DS_L = (M, (g^t)_{t \in T})$ be a dynamical system on a monoid $L = (T, +)$.

DS_L is complex := DS_L is not simple.

4 Past and Future

Let $DS_L = (M, (g^v)_{v \in T})$ be a dynamical system on a monoid $L = (T, +)$. The family of state transitions $(g^v)_{v \in T}$ allows us to introduce purely dynamical concepts of *past* and *future* as follows. Let 0 be the unity of L, $t \in T - \{0\}$, and $x \in M$.

Definition 7. *The t-past of* $x := P^t(x) := \{y : y \in M \text{ and } g^t(y) = x\}$.

Definition 8. *The t-future of* $x := F^t(x) := \{y : y \in M \text{ and } g^t(x) = y\}$.

Definition 9. *The past of* $x := P(x) := \bigcup_{t \in T - \{0\}} P^t(x)$.

Definition 10. *The future of* $x := F(x) := \bigcup_{t \in T - \{0\}} F^t(x)$.

Analogous definitions can be given for a set of states $X \subseteq M$. Let 0 be the unity of the time model $L = (T, +)$, and $t \in T - \{0\}$.

Definition 11. *The t-past of* $X := P^t(X) := \{y : y \in M \text{ and } \exists x \in X \text{ such that } g^t(y) = x\}$.

Definition 12. *The t-future of* $X := F^t(X) := \{y : y \in M \text{ and } \exists x \in X \text{ such that } g^t(x) = y\}$.

Definition 13. *The past of* $X := P(X) := \bigcup_{t \in T - \{0\}} P^t(X)$.

Definition 14. *The future of* $X := F(X) := \bigcup_{t \in T - \{0\}} F^t(X)$.

5 Constituent Subspaces and the Decomposition Theorem

By Definitions 14 and 3, any subspace X of the state space M of a DS_L on L contains its future $F(X)$. However, X may not contain its past $P(X)$, as the following example shows.

Example 3. Let $DS_L = (Z, (s^t)_{t \in Z^+})$, where Z are the integers, Z^+ the non-negative integers and, for any $t \in Z^+$, s^t is the t-th iteration of the successor function on Z. Then, in the first place, DS_L is a dynamical system on $L = (Z^+, +)$. Furthermore, Z^+ is a subspace of Z; however, $P(Z^+) \nsubseteq Z^+$.

Whenever X also contains $P(X)$, X is called temporally complete.

Definition 15. Let $DS_L = (M, (g^t)_{t \in T})$ be a dynamical system on a monoid $L = (T, +)$.

X is a temporally complete subspace of M := X is a subspace of M and $P(X) \subseteq X$.

Definition 16. Let $DS_L = (M, (g^t)_{t \in T})$ be a dynamical system on a monoid $L = (T, +)$, $x_1, x_2 \in M$, and $n \in Z^+ - \{0\}$.

For any $n \in Z^+ - \{0\}$, the ternary relation x_1 *is temporally n-connected with* x_2 *in M* is recursively defined below.

1. x_1 is temporally 1-connected with x_2 in M iff $x_1 \neq x_2$ and $\exists t \in T$ such that $g^t(x_1) = x_2$ or $g^t(x_2) = x_1$;
2. x_1 is temporally $(n+1)$-connected with x_2 in M iff $x_1 \neq x_2$ and $\exists x_3 \in M$ such that $x_3 \neq x_1$, $x_3 \neq x_2$, x_1 is temporally n-connected with x_3 in M and x_3 is temporally 1-connected with x_2 in M.

Lemma 1 (ATTRACTION LEMMA) *Let $DS_L = (M, (g^t)_{t\in T})$ be a dynamical system on a monoid $L = (T, +)$, and $x^* \in M$.*

If X is a temporally complete subspace of M, $x \in X$, and x is temporally n-connected with x^* in M, then $x^* \in X$.

Proof. By induction on $n \in Z^+ - \{0\}$. The base of the induction follows from 1 of Definition 16 and from Definitions 15, 14, 13, 12, and 11. The step of the induction is a straightforward consequence of 2 of Definition 16. □

Proposition 3 *Let $DS_L = (M, (g^t)_{t\in T})$ be a dynamical system on a monoid $L = (T, +)$, and $x^* \in M$. Let X be a temporally complete subspace of M, and $DS_{X_L} = (X, (g^t/X)_{t\in T})$.*

If $x \in X$, and x is temporally n-connected with x^* in M, then x is temporally n-connected with x^* in X.

Proof. By induction on $n \in Z^+ - \{0\}$. The base of the induction follows from the definition of temporal n-connectedness (1 of Definition 16), Proposition 1, and the Attraction Lemma (Lemma 1). The step of the induction is a direct consequence of the definition of temporal n-connectedness (2 of Definition 16). □

Proposition 4 *Let $DS_L = (M, (g^t)_{t\in T})$ be a dynamical system on a monoid $L = (T, +)$, $x_1, x_2 \in M$, and $n \in Z^+ - \{0\}$.*

If x_1 is temporally n-connected with x_2 in M, then x_2 is temporally n-connected with x_1 in M.

Proof. By induction on $n \in Z^+ - \{0\}$. The base of the induction immediately follows from 1 of Definition 16. The step of the induction easily follows from 2 of Definition 16. □

Definition 17. Let $DS_L = (M, (g^t)_{t\in T})$ be a dynamical system on a monoid $L = (T, +)$, and $x_1, x_2 \in M$.

x_1 is temporally connected with x_2 in M := $x_1 \neq x_2$ and $\exists n \in Z^+ - \{0\}$ such that x_1 is temporally n-connected with x_2 in M.

Definition 18. Let $DS_L = (M, (g^t)_{t\in T})$ be a dynamical system on a monoid $L = (T, +)$.

X is temporally connected in M := $X \subseteq M$ and, $\forall x_1, x_2 \in X$, if $x_1 \neq x_2$, then x_1 is temporally connected with x_2 in M.

Definition 19. Let $DS_L = (M, (g^t)_{t\in T})$ be a dynamical system on a monoid $L = (T, +)$.

X is a constituent subspace of M := X is a temporally complete subspace of M and X is temporally connected in M.

Definition 20. Let $DS_L = (M, (g^t)_{t \in T})$ be a dynamical system on a monoid $L = (T, +)$.

The decomposition of M into its constituent subspaces $:= \mathbf{C}_M := \{X : X$ is a constituent subspace of $M\}$.

In order to prove the *Decomposition Theorem* (Theorem 1 below), we need three more definitions and the *x-Constituent Lemma* (Lemma 2 below).

Definition 21. Let $DS_L = (M, (g^t)_{t \in T})$ be a dynamical system on a monoid $L = (T, +)$, $x \in M$, and $n \in Z^+$.

For any $n \in Z^+$, the set $X_x^n \subseteq M$ is recursively defined below.

1. $X_x^0 = \{x\}$;
2. $X_x^{n+1} = P(X_x^n) \cup X_x^n \cup F(X_x^n)$.

Definition 22. Let $DS_L = (M, (g^t)_{t \in T})$ be a dynamical system on a monoid $L = (T, +)$, and $x \in M$.

$X_x := \bigcup_{n \in Z^+} X_x^n$; X_x is called the *x-constituent* of M.

Note that, by Definition 22, $\forall z \in X_x, \exists n \in Z^+$ such that $z \in X_x^n$. We then define:

Definition 23. Let $DS_L = (M, (g^t)_{t \in T})$ be a dynamical system on a monoid $L = (T, +)$, $x \in M$, and $z \in X_x$.

The level of $z := lev(z) :=$ the minimum n such that $z \in X_x^n$.

Note that, by Definitions 23, 22, and by 2 of Definition 21, $\forall z \in X_x$, $\forall m \in Z^+$, if $m \geq lev(z)$, then $z \in X_x^m$.

Lemma 2 (*x*-**Constituent Lemma**) *Let $DS_L = (M, (g^t)_{t \in T})$ be a dynamical system on a monoid $L = (T, +)$, and $x \in M$.*

X_x is a constituent subspace of M, that is to say,

1. *X_x is a subspace of M;*
2. *X_x is a temporally complete subspace of M;*
3. *X_x is temporally connected in M.*

Proof. We prove the three theses below.

1. **Thesis 1:** X_x is a subspace of M.
2. By the definitions of subspace (Definition 3) and $F(z)$ (Definition 10), it suffice to prove: $\forall z \in X_x, F(z) \subseteq X_x$;
3. suppose: $lev(z) = n \in Z^+$;
4. by 3 and the definition of $lev(z)$ (Definition 23): $z \in X_x^n$;
5. by 4 and by the definitions of $F(z)$ and $F(X_x^n)$ (Definition 10 and 14): $F(z) \subseteq F(X_x^n)$;
6. by 5 and by the definitions of X_x^{n+1} and X_x (2 of Definition 21 and Definition 22): $F(z) \subseteq F(X_x^n) \subseteq X_x^{n+1} \subseteq X_x$.

1. **Thesis 2:** X_x is a temporally complete subspace of M.
2. By the definitions of temporally complete subspace (Definition 15), by thesis 1 and by the definitions of $P(X_x)$ and $P(z)$ (Definitions 13 and 9), it suffice to prove: $\forall z \in X_x, P(z) \subseteq X_x$;
3. suppose: $lev(z) = n \in Z^+$;
4. by 3 and the definition of $lev(z)$ (Definition 23): $z \in X_x^n$;
5. by 4 and by the definitions of $P(z)$ and $P(X_x^n)$ (Definitions 9 and 13): $P(z) \subseteq P(X_x^n)$;
6. by 5 and by the definitions of X_x^{n+1} and X_x (2 of Definition 21 and Definition 22): $P(z) \subseteq P(X_x^n) \subseteq X_x^{n+1} \subseteq X_x$.

1. **Thesis 3:** X_x is temporally connected in M.
2. By the definition of temporally connected subset (Definition 18) and by the definition of the relationship of temporal connectedness (Definition 17), theses 3 is equivalent to:
   for any $x_1, x_2 \in X_x$, if $x_1 \neq x_2$, then $\exists n \in Z^+ - \{0\}$ such that x_1 is temporally n-connected with x_2 in M;
3. we prove 2 by double induction, first on the level of x_2, and second on the level of x_1;

   a. **Base of the induction.** Suppose: $lev(x_1) = 0$ and $lev(x_2) = 0$;
   b. by 3a and by the definitions of level and X_x^0 (Definitions 23 and 1 of Definition 21): $x_1, x_2 \in X_x^0 = \{x\}$;
   c. by 3b: $x_1 = x_2$;
   d. thus, by 3c, 2 is vacuously satisfied.

   a. **Step of the induction on the level of** x_2. Suppose: $lev(x_1) = 0$, and 2 holds for any x_2 such that $lev(x_2) \leq m \in Z^+$;
   b. by 3a: $x_1 = x$;
   c. by 3b and 3a, it suffice to prove: for any $z_2 \in X_x$, if $lev(z_2) = m+1$ and $x \neq z_2$, then $\exists k \in Z^+ - \{0\}$ such that x is temporally k-connected with z_2 in M;
   d. suppose: $lev(z_2) = m+1$ and $x \neq z_2$;
   e. by 3d: $z_2 \in X_x^{m+1} = P(X_x^m) \cup X_x^m \cup F(X_x^m)$;
   f. by 3e and 3d: $z_2 \in P(X_x^m)$ or $z_2 \in F(X_x^m)$, but $z_2 \notin X_x^m$;
   g. by 3f: $\exists x_2^* \in X_x^m, \exists t \in T - \{0\}$ such that $g^t(z_2) = x_2^*$ or $g^t(x_2^*) = z_2$;
   h. by 3g and by the definition of level (Definition 23): $lev(x_2^*) \leq m$;
   i. by 3h, 3a, and 3b: if $x \neq x_2^*$, $\exists n \in Z^+ - \{0\}$ such that x is temporally n-connected with x_2^* in M;
   j. case: $x = x_2^*$;
      i. by 3j and 3g: $\exists t \in T - \{0\}$ such that $g^t(z_2) = x$ or $g^t(x) = z_2$;
      ii. by 3(j)i, 3d, and by the definition of temporal n-connectedness (1 of Definition 16): x is temporally 1-connected with z_2 in M; [3c is thus proved for case 3j]
   k. case: $x \neq x_2^*$;
      i. by 3f and 3g: $z_2 \neq x_2^*$;

ii. by 3k and 3i: $\exists n \in Z^+ - \{0\}$ such that x is temporally n-connected with x_2^* in M;
iii. by 3(k)i, 3g, and by the definition of temporal n-connectedness (1 of Definition 16): x_2^* is temporally 1-connected with z_2 in M;
iv. by 3(k)iii, 3(k)ii, 3d, and by the definition of temporal n-connectedness (2 of Definition 16): x is temporally $(n+1)$-connected with z_2 in M. [3c is thus proved]

a. **Step of the induction on the level of x_1.** Suppose: $lev(x_2) = r \in Z^+$, and 2 holds for any x_1 such that $lev(x_1) \leq m \in Z^+$;
b. by 3a, it suffice to prove: for any $z_1 \in X_x$, if $lev(z_1) = m+1$ and $z_1 \neq x_2$, then $\exists k \in Z^+ - \{0\}$ such that z_1 is temporally k-connected with x_2 in M;
c. suppose: $lev(z_1) = m+1$ and $z_1 \neq x_2$;
d. by 3c: $z_1 \in X_x^{m+1} = P(X_x^m) \cup X_x^m \cup F(X_x^m)$;
e. by 3d and 3c: $z_1 \in P(X_x^m)$ or $z_1 \in F(X_x^m)$, but $z_1 \notin X_x^m$;
f. by 3e: $\exists x_1^* \in X_x^m, \exists t \in T - \{0\}$ such that $g^t(z_1) = x_1^*$ or $g^t(x_1^*) = z_1$;
g. by 3f and by the definition of level (Definition 23): $lev(x_1^*) \leq m$;
h. by 3g and 3a: if $x_1^* \neq x_2$, $\exists n \in Z^+ - \{0\}$ such that x_1^* is temporally n-connected with x_2 in M;
i. case: $x_1^* = x_2$;
   i. by 3i and 3f: $\exists t \in T - \{0\}$ such that $g^t(z_1) = x_2$ or $g^t(x_2) = z_1$;
   ii. by 3(i)i, 3c, and by the definition of temporal n-connectedness (1 of Definition 16): z_1 is temporally 1-connected with x_2 in M; [3b is thus proved for case 3i]
j. case: $x_1^* \neq x_2$;
   i. by 3e and 3f: $z_1 \neq x_1^*$;
   ii. by 3j and 3h: $\exists n \in Z^+ - \{0\}$ such that x_1^* is temporally n-connected with x_2 in M;
   iii. by 3(j)i, 3f, and by the definition of temporal n-connectedness (1 of Definition 16): x_1^* is temporally 1-connected with z_1 in M;
   iv. by 3(j)iii, 3(j)ii, 3c, by the definition of temporal n-connectedness (2 of Definition 16), and by commutativity of temporal n-connectedness (Proposition 4, applied twice): z_1 is temporally $(n+1)$-connected with x_2 in M. [3b is thus proved]

□

Theorem 1 (DECOMPOSITION THEOREM) *Let $DS_L = (M, (g^t)_{t \in T})$ be a dynamical system on a monoid $L = (T, +)$, and $\mathbf{C}_M$ be the decomposition of M into its constituent subspaces.*

$\mathbf{C}_M$ is a partition of M, that is to say,

1. *for any $X, Z \in \mathbf{C}_M$, if $X \neq Z$, then $X \cap Z = \emptyset$;*
2. $\bigcup_{X \in \mathbf{C}_M} X = M$.

Proof. We prove the two theses below.

1. **Thesis 1:** for any $X, Z \in \mathbf{C}_M$, if $X \neq Z$, then $X \cap Z = \emptyset$.
2. Suppose for *reductio*: $\exists X, Z \in \mathbf{C}_M : X \neq Z$ and $X \cap Z \neq \emptyset$;
3. we show below:

    a. $Z \subseteq X$;
    b. $X \subseteq Z$;
        i. **Thesis 3a:** $Z \subseteq X$.
        ii. Suppose for *reductio*: $Z \nsubseteq X$;
        iii. by 3(b)ii: $\exists z : z \in Z$ and $z \notin X$;
        iv. as, by 2, $X \cap Z \neq \emptyset$, suppose: $z^* \in X \cap Z$;
        v. by 2: Z is temporally connected in M;
        vi. by 3(b)iv and 3(b)iii: $z \neq z^*$;
        vii. by 3(b)vi, 3(b)v, 3(b)iv, 3(b)iii and by the definitions of temporally connected subset (Definition 18), temporal connectedness (Definition 17) and temporal n-connectedness (Definition 16): there is $n \in Z^+ - 0$ such that z^* is temporally n-connected with z in M;
        viii. by 2: X is a temporally complete subspace of M;
        ix. by 3(b)viii, 3(b)vii, 3(b)iv, and by the Attraction Lemma (Lemma 1): $z \in X$, contrary to 3(b)iii.
        i. **Thesis 3b:** $X \subseteq Z$.
        ii. the proof of thesis 3b is completely analogous to the proof of thesis 3a.

4. by 3a and 3b: $X = Z$, contrary to 2.

1. **Thesis 2:** $\bigcup_{X \in \mathbf{C}_M} X = M$.
2. by the definitions of $\mathbf{C}_M$ (Definition 20), constituent subspace of M (Definition 19), temporally complete subspace of M (Definition 15), and subspace of M (Definition 3): $\bigcup_{X \in \mathbf{C}_M} X \subseteq M$;
3. by the definition of X_x (Definition 22), and by the x-Constituent Lemma (Lemma 2): $\forall x \in M, x \in X_x \in \mathbf{C}_M$;
4. by 3: $M \subseteq \bigcup_{X \in \mathbf{C}_M} X$;
5. by 4 and 2: $\bigcup_{X \in \mathbf{C}_M} X = M$.

□

Given a set X and a property Φ such that X has Φ, X *is minimal with respect to* Φ := there is no Y such that $Y \subset X$ and Y has Φ; X *is maximal with respect to* Φ := there is no Y such that $Y \supset X$ and Y has Φ. It is then immediate to show:

Corollary 1 *Let $DS_L = (M, (g^t)_{t \in T})$ be a dynamical system on a monoid $L = (T, +)$, X be a constituent subspace of M, and Φ be the property of being a constituent subspace of M.*

X is both minimal and maximal with respect to Φ.

Proof. The thesis is an immediate consequence of the definitions of minimality, maximality, Definition 20, and thesis 1 of Theorem 1. □

6 Constituent Subsystems

Definition 24. Let $DS_L = (M, (g^t)_{t \in T})$ be a dynamical system on a monoid $L = (T, +)$.

CS_{X_L} is a constituent subsystem of DS_L := $X \in \mathbf{C}_M$ and $CS_{X_L} = (X, (g^t/X)_{t \in T})$.

Proposition 5 *Let $DS_L = (M, (g^t)_{t \in T})$ be a dynamical system on a monoid $L = (T, +)$.*

If CS_{X_L} is a constituent subsystem of DS_L, then CS_{X_L} is a subsystem of DS_L.

Proof. Suppose CS_{X_L} is a constituent subsystem of DS_L. Thus, by Definitions 24, 20, 19, and 15, X is a subspace of M. Hence, by Definition 24 and Proposition 2, CS_{X_L} is a subsystem of DS_L. □

Definition 25. Let $DS_L = (M, (g^t)_{t \in T})$ be a dynamical system on a monoid $L = (T, +)$.

The decomposition of DS_L into its constituent subsystems := $(CS_{X_L})_{X \in \mathbf{C}_M}$.

Note that a constituent subsystem of a dynamical system DS_L on a monoid L may very well be complex, as the following example shows.

Example 4. Let us consider the dynamical system of Example 3. Recall that $DS_L = (Z, (s^t)_{t \in Z^+})$ is a dynamical system on $(Z^+, +)$, where s^t is the t-th iteration of the successor function on Z. In fact, the only constituent subsystem of DS_L is DS_L itself; however, $(Z^+, (s^t/Z^+)_{t \in Z^+}) \neq DS_L$ is a subsystem of DS_L. Thus, DS_L is not simple, that is to say, DS_L is complex.

Definition 26. Let $DS_L = (M, (g^t)_{t \in T})$ be a dynamical system on a monoid $L = (T, +)$.

DS_L is indecomposable := DS_L does not have any constituent subsystem but itself.

Definition 27. Let $DS_L = (M, (g^t)_{t \in T})$ be a dynamical system on a monoid $L = (T, +)$.

DS_L is decomposable := DS_L is not indecomposable.

Example 5. The first example below is a complex indecomposable system, while the second one is decomposable.

1. Let $DS1_L = (Z, (s^t)_{t \in Z^+})$, where s^t is the t-th iteration of the successor function on Z. $DS1_L$ is a dynamical system on $L = (Z^+, +)$, and $DS1_L$ is complex and indecomposable (see Example 4).
2. Let $DS2_L = (Z, (id^t)_{t \in Z^+})$, where id^t is the t-th iteration of the identity function on Z. $DS2_L$ is a dynamical system on $L = (Z^+, +)$, and $DS2_L$ is decomposable. (Hence, $DS2_L$ is complex as well.)

Proposition 6 *Let $DS_L = (M, (g^t)_{t \in T})$ be a dynamical system on a monoid $L = (T, +)$.*

If CS_{X_L} is a constituent subsystem of DS_L, then CS_{X_L} is indecomposable.

Proof. Suppose CS_{X_L} is a constituent subsystem of DS_L and, for *reductio*, that CS_{X_L} is decomposable. Let M_1 and M_2 be the state spaces of two distinct constituent subsystems of CS_{X_L}. Then, in the first place, M_1 is a temporally complete subspace of X. Let $x_1 \in M_1$ and $x_2 \in M_2$. As M_1 is a temporally complete subspace of X, by the Attraction Lemma (Lemma 1), thesis 1 of the Decomposition Theorem (Theorem 1), and by the definitions of temporal n-connectedness (Definition 17) and temporal connectedness (Definition 17), x_1 is not temporally connected with x_2 in X. On the other hand, as CS_{X_L} is a constituent subsystem of DS_L, X is temporally connected in M, and thus x_1 is temporally connected with x_2 in M. Furthermore, X is a temporally complete subspace of M; hence, by Proposition 3, x_1 is temporally connected with x_2 in X. Contradiction. □

7 Composition of Disjoint Dynamical Systems

Definition 28. Let $DS1_L = (M_1, (g_1^t)_{t \in T})$ and $DS2_L = (M_2, (g_2^t)_{t \in T})$ be two dynamical systems on the same monoid $L = (T, +)$.

$DS1_L$ and $DS2_L$ are disjoint $:= M_1 \cap M_2 = \emptyset$.

Proposition 7 *Let $DS_L = (M, (g^t)_{t \in T})$ be a dynamical system on a monoid $L = (T, +)$.*

If CS_{X_L} and CS_{Z_L} are distinct constituent subsystems of DS_L, then CS_{X_L} and CS_{Z_L} are disjoint.

Proof. By the definition of constituent subsystem (Definition 24), both X and Z are constituent subspaces of M. Therefore, by Theorem 1, $X \cap Z = \emptyset$; thus, by the definition of disjoint subsystems (Definition 28), CS_{X_L} and CS_{Z_L} are disjoint. □

Any two disjoint dynamical systems on the same monoid L can always be composed into a single dynamical system on L by means of the composition operation $\oplus$, which is defined below.

Definition 29. Let $DS1_L = (M_1, (g_1^t)_{t \in T})$ and $DS2_L = (M_2, (g_2^t)_{t \in T})$ be two disjoint dynamical systems on the same monoid $L = (T, +)$.

The composition of $DS1_L$ and $DS2_L$ $:= DS1_L \oplus DS2_L := (M_1 \cup M_2, (g_1^t \oplus g_2^t)_{t \in T})$ where, $\forall t \in T$, $\forall x \in M_1 \cup M_2$, $g_1^t \oplus g_2^t(x) := g_1^t(x)$, if $x \in M_1$; $g_1^t \oplus g_2^t(x) := g_2^t(x)$, if $x \in M_2$.

Proposition 8 *Let $DS1_L = (M_1, (g_1^t)_{t \in T})$ and $DS2_L = (M_2, (g_2^t)_{t \in T})$ be two disjoint dynamical systems on the same monoid $L = (T, +)$.*

1. *$DS1_L \oplus DS2_L$ is a dynamical system on L;*
2. *$\oplus$ is commutative.*

Proof. Thesis 1 easily follows from the definitions of composition (Definition 29) and dynamical system on a monoid (Definition 2). As for thesis 2, it is an immediate consequence of Definition 29. □

As shown below, the composition operation can be generalized to any family of mutually disjoint dynamical systems on a given monoid L.

Definition 30. Let $L = (T, +)$ be a monoid.

$(DS_{X_L})_{X\in\mathbf{D}}$ *is a family indexed by* $\mathbf{D}$ *of mutually disjoint dynamical systems on* $L := (DS_{X_L})_{X\in\mathbf{D}} = ((X, (g_X^t)_{t\in T}))_{X\in\mathbf{D}}$ is a family, indexed by $\mathbf{D}$, of dynamical systems on L, and $\mathbf{D}$ is a set of mutually disjoint sets (i.e., for any $X, Z \in \mathbf{D}$, if $X \neq Z$, then $X \cap Z = \emptyset$).

Let $(DS_{X_L})_{X\in\mathbf{D}} = ((X, (g_X^t)_{t\in T}))_{X\in\mathbf{D}}$ be a family, indexed by $\mathbf{D}$, of mutually disjoint dynamical systems on a given monoid $L = (T, +)$. As $\mathbf{D}$ is a set of mutually disjoint sets, $\mathbf{D}$ is a partition of $\bigcup_{X\in\mathbf{D}} X$. Consequently, $\forall x \in \bigcup_{X\in\mathbf{D}} X$, there is exactly one $X \in \mathbf{D}$ such that $x \in X$. Thus, let us define the function χ as follows:

Definition 31. $\chi : \bigcup_{X\in\mathbf{D}} X \to \mathbf{D}$, $\forall x \in \bigcup_{X\in\mathbf{D}} X$, $\chi(x) =$ the $X \in \mathbf{D}$ such that $x \in X$.

We can now generalize the composition operation as follows.

Definition 32. Let $(DS_{X_L})_{X\in\mathbf{D}} = ((X, (g_X^t)_{t\in T}))_{X\in\mathbf{D}}$ be a family, indexed by $\mathbf{D}$, of mutually disjoint dynamical systems on a given monoid $L = (T, +)$.

$$\sum_{X\in\mathbf{D}} DS_{X_L} = \Big(\bigcup_{X\in\mathbf{D}} X, \big(\sum_{X\in\mathbf{D}} g_X^t\big)_{t\in T}\Big), where$$

$$\forall t \in T, \forall x \in \bigcup_{X\in\mathbf{D}} X, \sum_{X\in\mathbf{D}} g_X^t(x) = g_{\chi(x)}^t(x).$$

Proposition 9 *Let* $(DS_{X_L})_{X\in\mathbf{D}} = ((X, (g_X^t)_{t\in T}))_{X\in\mathbf{D}}$ *be a family, indexed by* $\mathbf{D}$, *of mutually disjoint dynamical systems on a given monoid* $L = (T, +)$.

$\Sigma_{X\in\mathbf{D}} DS_{X_L}$ *is a dynamical system on L.*

Proof. The thesis follows from the definitions of generalized composition (Definition 32), χ function (Definition 31), and dynamical system on a monoid (Definition 2). □

8 The Composition Theorem

Let $DS_L = (M, (g^t)_{t\in T})$ be a dynamical system on a monoid $L = (T, +)$, and $(CS_{X_L})_{X\in\mathbf{C}_M} = ((X, (g^t/X)_{t\in T}))_{X\in\mathbf{C}_M}$ be the decomposition of DS_L into its constituent subsystems (see Definitions 25 and 24). By Definition 30, $(CS_{X_L})_{X\in\mathbf{C}_M}$ is a family, indexed by $\mathbf{C}_M$, of mutually disjoint dynamical systems on L. Thus, the generalized composition operation applies to $(CS_{X_L})_{X\in\mathbf{C}_M}$, and it holds:

Theorem 2 (Composition Theorem)

$$\Big(\sum_{X\in\mathbf{C}_M} CS_{X_L}\Big) = DS_L$$

Proof. The thesis is a straightforward consequence of the definition of generalized composition (Definition 32), the Decomposition Theorem (Theorem 1), and Proposition 9. □

References

1. Arnold, V. I. (1977). *Ordinary differential equations*. Cambridge, MA: The MIT Press.
2. Giunti, M. (1997). *Computation, dynamics, and cognition*. New York: Oxford University Press.
3. Giunti, M., & Mazzola, C. (2012). Dynamical systems on monoids: Toward a general theory of deterministic systems and motion. In G. Minati, M. Abram, & E. Pessa (Eds.), *Methods, models, simulations and approaches towards a general theory of change* (pp. 173–185). Singapore: World Scientific.
4. Hirsch, M. W., Smale, S., & Devaney, R. L. (2004). *Differential equations, dynamical systems, and an introduction to chaos*. Amsterdam: Elsevier Academic Press.
5. Szlenk, W. (1984). *An introduction to the theory of smooth dynamical systems*. New York: Wiley.
6. Wolfram, S. (1983). Cellular automata. *Los Alamos Science, 9*, 2–21.

Chapter 7
For a Topology of Dynamical Systems

Claudio Mazzola and Marco Giunti

1 Introduction

Dynamical systems are mathematical objects meant to formally capture the dynamical features of deterministic systems. They are commonly defined as ordered pairs of the form $DS = (S, (f^t)_{t \in T})$, where S is a non-empty set of *states* or *points* called the *state space*, and $(f^t)_{t \in T}$ is a family of functions on S, indexed by T, called *state transitions*. For every $t \in T$, the state transition f^t is said to have *duration* t, where the *time set* T is usually taken to be a set of numbers, such as the reals $\mathscr{R}$, the non-negative reals $\mathscr{R}^0$, the integers $\mathscr{Z}$, or the non-negative integers $\mathscr{Z}^0$. Each state transition specifies the way its argument evolves in the time given by the corresponding duration. More specifically, it is required that the state transition of duration 0 is the identity map on S, while the composition of any two state transitions is identical to the state transition whose duration is the sum of their durations [3, 4].

So defined, dynamical systems suffice to model an extensive class of deterministic systems, ranging from classical pendulums to cellular automata. Nonetheless, it is possible to further generalize their definition as follows:

C. Mazzola (✉)
School of History, Philosophy, Religion and Classics, The University of Queensland, Forgan Smith Building (1), Brisbane, QLD 4072, Australia
e-mail: c.mazzola@uq.edu.au

M. Giunti
ALOPHIS, Applied Logic Language Philosophy and History of Science, University of Cagliari, Via is Mirrionis 1, 09123 Cagliari, Italy
e-mail: giunti@unica.it

G. Minati et al. (eds.), *Towards a Post-Bertalanffy Systemics*, Contemporary Systems Thinking, DOI 10.1007/978-3-319-24391-7_7

Definition 7.1. Let S be a non-empty set, let $M = (T,+)$ be a monoid with identity 0 and let $(f^t)_{t\in T}$ be a family of functions on S indexed by T. The ordered pair $DS_M = (S,(f^t)_{t\in T})$ is a *dynamical system on M* if and only if:

$$\forall x \in S, \forall t,v \in T: \; f^0(x) = x, \tag{7.1}$$

$$f^{t+v}(x) = f^t(f^v(x)). \tag{7.2}$$

The monoid M is here called the *time model* of the dynamical system, while T is called its *time set*. Giunti and Mazzola have shown that dynamical systems on monoids adequately capture all the fundamental notions of dynamical system theory, whereas no poorer mathematical structure would equally do the job. Notably, it is not possible to further reduce the algebraic properties of time models without compromising the ability of dynamical systems to model deterministic change [2].

By contrast, it appears remarkable that no similar constraint is imposed on state spaces. Most notably, these are not required to possess any topological structure, despite the fact that the evolution of a dynamical system is often analysed by looking at the topological features of its diagrammatical representation. To wit, a system whose evolution exhibits some periodicity can be represented by means of a closed line, while the evolution of a completely aperiodic system is modelled by an open curve. Two intersecting lines represent a system in which distinct inputs deliver the same output, and so on.

Similar considerations suggest that it is possible to recover the topological properties of state spaces from the dynamical properties of the corresponding systems, and possibly that any such topological property might supervene on some dynamical feature. The objective of this article is to make a preliminary step in this field, by examining whether some workable topology can actually be defined within the formal language of dynamical systems theory.

2 Outward Topologies

The following notions will be especially useful to us.

Definition 7.2. Let $DS_M = (S,(f^t)_{t\in T})$ be a dynamical system on $M = (T,+)$. For any $x \in S$, the *orbit* of x is the set:

$$O(x) := \{y \in S : \exists t \in T (y = f^t(x))\}. \tag{7.3}$$

Definition 7.3. Let $DS_M = (S,(f^t)_{t\in T})$ be a dynamical system on $M = (T,+)$. For any $x \in S$ and any $t \in T$, the *ϕ-interval* of x of duration t is the set of all states in the orbit of x from which $f^t(x)$ can be reached:

$$\phi_t(x) := \{y \in S : y \in O(x) \wedge f^t(x) \in O(y)\}, \tag{7.4}$$

Definition 7.4. Let $DS_{M_1} = (S_1, (f^{t_1})_{t_1 \in T_1})$ be a dynamical system on a monoid $M_1 = (T_1, +)$, let $DS_{M_2} = (S_2, (f^{t_2})_{t_2 \in T_2})$ be a dynamical system on a monoid $M_2 = (T_2, \oplus)$, and let $\rho : T_1 \to T_2$ be a monoid isomorphism. The function $\psi : S_1 \to S_2$ is a ρ*-emulation* of DS_{M_1} in DS_{M_2} if and only if it is injective and

$$\forall x_1 \in S_1, \forall t_1 \in T_1 \ \ (f^{\rho(t_1)}(\psi(x_1)) = \psi(f^{t_1}(x_1))). \tag{7.5}$$

The orbit of a point x intuitively encompasses all the states the system will eventually evolve into if initially set in that state, while the ϕ-interval of duration t of x is the subset of points in $O(x)$ from which $f^t(x)$ can be reached. The notion of ρ-emulation, instead, generalises the more common notion of emulation [1] to the case of dynamical systems on monoids with different time sets. We say that a dynamical system DS_{M_1} *emulates* a dynamical system DS_{M_2} just in case there is a ρ-emulation of the former into the latter. If that happens, the dynamics of DS_{M_1} is perfectly reproduced by DS_{M_2}, to the effect that for any state in DS_{M_1} there is a state in DS_{M_2} whose orbit has the same dynamical properties as the orbit of the former.

The easiest way to define a topology using the minimal vocabulary just provided is to simply take a neighborhood of a point to be a superset of its orbit, accordingly identifying open sets with sets that contain the orbits of all their elements. The class of open sets so obtained, supplemented with the empty set, could then be easily proved to generate a topology on the state space of the system, which we might call its *orbit topology*. Notably, orbit topologies are preserved by ρ-emulation, in the sense that every ρ-emulation of a dynamical system into another is a continuous function from the orbit topology of the former to the orbit topology of the latter.

This result is surely of interest, since it demonstrates that it is always possible to define a topology on the state space of a dynamical system starting from its sole dynamical properties. Furthermore, such a topology supervenes on the dynamical features of the system, to the effect that two dynamical systems cannot differ as to the former without exhibiting different dynamics. Nonetheless, orbit topologies have their downsides. To wit, owing to the fact that $x \in O(y)$ just in case $O(x) \subseteq O(y)$, no orbit topology can be Hausdorff, unless all points in the dynamical system are fixed points.

Our task will be, accordingly, to find a more appropriate way of identifying neighborhoods and open sets on the state space of a dynamical system. Let us begin, then, with the notion of open sets. We may try and refine the intuition underlying the orbit topology by requiring that an open set should include some ϕ-interval for any one of its elements. The class of sets satisfying this condition demonstrably generates a topology; however, that topology is of scarce interest, because it amounts to the discrete topology on the state space. A better option is obtained by excluding the ϕ-interval of null duration:

Definition 7.5. Let $DS_M = (S, (f^t)_{t \in T})$ be a dynamical system on a monoid $M = (T, +)$ with identity 0. For any $X \subseteq S$, X is *outward open* (or *o-open*) if and only if:

$$\forall x \in X, \exists t \neq 0 \in T \, (\phi_t(x) \subseteq X). \tag{7.6}$$

Correspondingly, we define an *o-neighborhood* of a point as a superset of an o-open set including that point.

How is Definition 7.5 meant to provide a dynamical counterpart of the idea of open set? Notice that, for every point x in an o-open set X, there is a ϕ-interval $\phi_t(x)$ of non-null duration that is contained in X. Hence, there is no greatest ϕ-interval of x that is included in X. In plain words this means that, moving along the orbit of x, there is no last point one encounters before leaving X, i.e. before reaching a point in the orbit of x that does not belong to X. This is in perfect agreement with the intuition that an open set is one that does not include its own boundary.

To formally confirm the adequacy of the above definition, let us define:

Definition 7.6. Let $DS_M = (S, (f^t)_{t\in T})$ be a dynamical system on $M = (T, +)$. DS_M is *ϕ-linear* if and only if

$$\forall x \in S, \forall t, v \in T (\phi_t(x) \subseteq \phi_v(x) \vee \phi_v(x) \subseteq \phi_t(x)). \qquad (7.7)$$

This condition amounts to the request that $\subseteq$ be a linear order on the set of the ϕ-intervals of each point x. It is satisfied by all dynamical systems whose time model is a numerical set, along with the arithmetic operation of addition. It is therefore a plausible and relatively undemanding constraint to impose on a dynamical system. We can thus prove that:

Proposition 7.1. *Let $DS_M = (S, (f^t)_{t\in T})$ be a ϕ-linear dynamical system on a monoid $M = (T, +)$ with identity 0 and let Σ_o be the set of the o-open subsets of S. Then (S, Σ_o) is a topological space.*

Proof. To prove that (S, Σ_o) is a topological space, we need to show: (1) that S and $\emptyset$ are o-open; (2) that the union of any given collection of o-open subsets of S is o-open; and (3) that the intersection of any two o-open subsets of S is o-open.

To prove (1), it is sufficient to notice that S satisfies (7.6) by definition of dynamical system on a monoid, while $\emptyset$ satisfies it vacuously.

To prove (2), let $\Gamma \subseteq \Sigma_o$ be any collection of o-open subsets of S, and let Θ be the union of all the elements of Γ. If $\Theta = \emptyset$, then it is o-open, as just demonstrated. Otherwise, let $x \in \Theta$. By definition of Θ, there must exist some o-open set $X \in \Gamma$ such that $x \in X$. By (7.6) there is in consequence some $t \neq 0 \in T$ such that $\phi_t(x) \subseteq X \subseteq \Theta$. Since x was chosen arbitrarily among the elements of Θ, this shows that Θ is o-open.

To prove (3), let $X_1, X_2 \in \Sigma_o$. If $X_1 \cap X_2 = \emptyset$, then it is o-open. Otherwise, let $x \in X_1 \cap X_2$. Then, by (7.6) there exist $t \neq 0, v \neq 0 \in T$ such that $\phi_t(x) \subseteq X_1$ and $\phi_v(x) \subseteq X_2$. By ϕ-linearity, $\phi_t(x) \subseteq \phi_v$ or $\phi_v(x) \subseteq \phi_t$, so either $\phi_t(x) \subseteq X_1 \cap X_2$ or $\phi_v(x) \subseteq X_1 \cap X_2$. In either case, $X_1 \cap X_2$ is o-open. □

Let us call the topology so obtained the *outward topology* or *o-topology* of DS_M. It can be demonstrated that ϕ-linearity is a sufficient but not necessary condition for Σ_o to be an o-topology. However, dynamical systems that are not ϕ-linear are likely to lack any straightforward interpretation; therefore we can hereafter restrict our attention to ϕ-linear dynamical systems without any significant loss of generality.

It is also interesting to notice that every orbit is an o-open set, and that in consequence the o-topology of a dynamical system is identical to or finer than its orbit topology. Moreover, just like the latter, the o-topology is preserved by ρ-emulation.

Proposition 7.2. *Let $DS_{M_1} = (S_1, (f^{t_1})_{t_1 \in T_1})$ be a ϕ-linear dynamical system on a monoid $M_1 = (T_1, +)$ with identity 0, let $DS_{M_2} = (S_2, (f^{t_2})_{t_2 \in T_2})$ be a ϕ-linear dynamical system on a monoid $M_2 = (T_2, \oplus)$ with identity θ, and let $\rho : T_1 \to T_2$ be a monoid isomorphism. Let Σ_{o_1} be the collection of o-open sets of DS_{M_1} and let Σ_{o_2} be the collection of o-open sets of DS_{M_2}. Then every ρ-emulation of DS_{M_1} in DS_{M_2} is a continuous function of (S_1, Σ_{o_1}) in (S_2, Σ_{o_2}).*

Proof. Let ψ be a ρ-emulation of DS_{M_1} in DS_{M_2}, and let $X_2 \subseteq S_2$ be an arbitrary o-open subset of S_2. Because ψ is injective, ψ^{-1} is well-defined. So, let $X_1 = \psi^{-1}(X_2)$. To show that ψ is a continuous function, it will be sufficient to show that for any $x_1 \in X_1$ there exists some $t_1 \neq 0 \in T_1$ such that any element of $\phi_{t_1}(x_1)$ is in X_1, which means that any such element is the counterimage of some element of X_2 with respect to ψ.

So, let $x_1 \in X_1$ be chosen arbitrarily and let $x_2 = \psi(x_1)$. Because X_2 is open, there exists $t_2 \neq \theta \in T_2$ such that $\phi_{t_2}(x_2) \subseteq X_2$. For any $x_{i_2} \in \phi_{t_2}(x_2)$, it is then clear that $x_{i_2} \in X_2$ and therefore, by hypothesis, $\psi^{-1}(x_{i_2}) \in X_1$. Notably, $\psi^{-1}(f^{t_2}(x_2)) \in X_1$. Furthermore, since ρ is a monoid isomorphism, there exists exactly one $t_1 \neq 0 \in T_1$ such that $t_2 = \rho(t_1)$. Hence:

$$\psi^{-1}(f^{t_2}(x_2)) = \psi^{-1}(f^{\rho(t_1)}(\psi(x_1))) = \psi^{-1}(\psi(f^{t_1}(x_1)) = f^{t_1}(x_1). \tag{7.8}$$

Take now any $x_{i_1} \in S_1$ such that $x_{i_1} \in \phi_{t_1}(x_1)$. Clearly, there must exist $t_{i_1}, t_{j_1} \in T_1$ such that $f^{t_{i_1}}(x_1) = (x_{i_1})$ and $f^{t_{j_1}}(x_{i_1}) = f^{t_1}(x_1)$. Therefore, by (7.5):

$$\psi(x_{i_1}) = \psi(f^{t_{i_1}}(x_1)) = \psi(f^{t_{i_1}}(\psi^{-1}(x_2))) = f^{\rho(t_{i_1})}(\psi(\psi^{-1}(x_2))) = f^{\rho(t_{i_1})}(x_2), \tag{7.9}$$

and

$$f^{t_2}(x_2) = \psi(\psi^{-1}(f^{t_2}(x_2))) = \psi(f^{t_1}(x_1)) = f^{\rho(t_{j_1})}(\psi(x_{i_1})). \tag{7.10}$$

From (7.9), it follows that $\psi(x_{i_1}) \in O(x_2)$, while (7.10) entails that $f^{t_2}(x_2) \in O(\psi(x_{i_1}))$. Therefore, $\psi(x_{i_1}) \in \phi_{t_2}(x_2) \subseteq X_2$ and thus $x_{i_1} \in X_1$. □

3 Dynamical Topologies

O-topologies share the virtues of orbit topologies without suffering from analogous shortcomings. They are, accordingly, better candidates to examine the relation between the dynamical and the topological features of a system. Nonetheless, they are not without defects. Consider, for instance, the dynamical system $DS_M = (S, (f^t)_{t \in T})$ on $M = (T, +)$, where $S = \mathscr{R}$ is the set of the real numbers, $T = \mathscr{R}^0$

is the set of the non-negative real numbers, $+$ is the standard addition operation and, for any $t \in T$ and any $x \in S$, $f^t(x) = t + x$. It is easy to see that the set $X_1 = \{x \in S : x < 0\}$ is an o-open set, and so it is $X_2 = \{x \in S : 0 \leq x\}$. But since $X_1 \cap X_2 = \emptyset$ and $X_1 \cup X_2 = \mathscr{R}$, it follows that $\mathscr{R}$ is disconnected, which is not what we would expect from the set of real numbers. For this reason, we need to identify a proper subclass of o-open sets, such that the topologies generated by that class avoid unpalatable consequences like the one just pointed out.

As we have noticed, o-open sets formally capture the idea that open sets are essentially unbounded, in the sense that there is no last point one needs to cross in order to step out of an open set; notably, this intuition is regimented through the requirement that, for any point in an o-open set, there is no greatest ϕ-interval belonging to that set. However, open sets are also intuitively unbounded in another sense, namely that there is no *first* point one encounters in an open set while stepping *inside* it. So, it seems that o-open sets only tell half of the dynamical story about open sets. The second half can be told with the aid of the following definitions:

Definition 7.7. Let $DS_M = (S, (f^t)_{t \in T})$ be a dynamical system on $M = (T, +)$. For any $x \in S$ and any $t \in T$, the *β-interval* of x of duration t is the set of all states whose orbits contain x, and that can be reached from some state z such that $f^t(z) = x$:

$$\beta_t(x) := \{y \in S : x \in O(y) \wedge \exists z \in S(x = f^t(z) \wedge y \in O(z))\}. \tag{7.11}$$

Definition 7.8. Let $DS_M = (S, (f^t)_{t \in T})$ be a dynamical system on a monoid $M = (T, +)$ with identity 0. For any $X \subseteq S$, X is *inward open* (or *i-open*) if and only if [1]:

$$\forall x \in X(\exists t \neq 0 \in T, \exists y \in S(f^t(y) = x) \rightarrow \exists v \neq 0 \in T(\emptyset \neq \beta_v(x) \subseteq X)). \tag{7.12}$$

Definition 7.9. Let $DS_M = (S, (f^t)_{t \in T})$ be a dynamical system on $M = (T, +)$. DS_M is *β-linear* if and only if

$$\forall x \in S, \forall t, v \in T(\beta_t(x) \subseteq \beta_v(x) \vee \beta_v(x) \subseteq \beta_t(x)). \tag{7.13}$$

It is clear that the notions of β-interval, i-open set and β-linearity are the duals of the notions of ϕ-interval, o-open set, and ϕ-linearity, respectively. It is thus no surprise that the class Σ_i of i-open sets of a β-linear dynamical system generates a topology on its state space, which we may call *inward topology*, and that all dynamical systems with numeric time models are β-linear. Furthermore, every ρ-emulation of a β-linear dynamical system into another is a continuous function from the inward topology of the former to the one of the latter. Proofs are similar to those of Propositions 7.1 and 7.2, respectively.

The o-open sets we are looking for are thus precisely the ones which are also i-open. We may call them *dynamically open*, or *d-open*. Given the above results, it

---

[1] Notice that, while for any $x \in X$ and any $t \neq 0 \in T$, the existence of $y = f^t(x) \in S$ is guaranteed by the definition of a dynamical system on a monoid, there is no similar guarantee that some $z \in S$ exists, for which $f^t(z) = x$. This explains why condition (7.12) below is comparatively stronger than the corresponding condition (7.6).

is elementary to prove that they generate a topology on every dynamical system that is both ϕ-linear and β-linear, and that such a topology is preserved by ρ-emulation. Let us label any such topology *dynamical*, or *d-topology*. It is immediate to see that d-topologies do not suffer from the type of shortcomings we saw to affect o-topologies. Most notably, the d-topology of the dynamical system considered at the beginning of this section is homeomorphic to the ordinary topology on the real numbers. It is thus reasonable to expect that d-open sets could adequately support a general examination of the way the dynamical features of a deterministic system naturally induce a topology on its state space.

4 Conclusion

Although no topological constraint is usually imposed on the state space of a dynamical system, there is prima facie evidence that its topological properties might naturally depend on the dynamical features of the system. This article has prepared the grounds for a systematic investigation of such dependence, by identifying in d-open sets promising candidates for the notion of a topology naturally induced by the underlying dynamics.

References

1. Giunti, M. (1997). *Computation, dynamics, and cognition*. New York: Oxford University Press.
2. Giunti, M., & Mazzola, C. (2012). Dynamical systems on monoids. Toward a general theory of deterministic systems and motion. In G. Minati, M. Abram, & E. Pessa (Eds.), *Models, simulations and approaches towards a general theory of change* (pp. 173–186). Singapore: World Scientific.
3. Hirsch, M. W., Smale, S., & Devaney, R. L. (2004). *Differential equations, dynamical systems, and an introduction to chaos*. Amsterdam: Elsevier.
4. Szlenk, W. (1984). *An introduction to the theory of smooth dynamical systems*. New York: Wiley.

Part IV
New Systemic Contents of Disciplinary Approaches and Problems

Chapter 8
EPAS: Artificial Intelligent System for Assistance

Guido Tascini

1 Introduction

From a general point of view, we define Artificial Intelligent Systems (AIS) machines that behave rationally, according to what says Artificial Intelligence. And we call rational behavior that of a machine, which uses computational models to solve problems. AIS are often Software Systems, in which intelligent machines are computational machines. The Physical AIS are instead those with the ability to perceive, move and act independently, pursuing own purposes and implementing plans to achieve them: they are physical machines that we call Robot.

2 Complex Intelligent Systems

In the meantime gradually are emerging Complex Intelligent Systems, using computational reasoning, able to learn and experience the world through sensors and actuators. Figure 8.1 shows the block diagram of the intelligent behavior of such systems.

The Artificial Intelligent Systems, after a start mostly of industrial applications, begin to affect the life of every day and promise a revolution in the way of life of everyone. The proof is the fact that many laboratories around the world have begun to study and implement AIS capable of supporting the man in his work, household chores, in everyday life, trying to make the machine as close as possible to man in terms of interaction.

The present work introduces the problem of AIS, able to provide a service as close as possible to a human service. Then describes a design of human-oriented

G. Tascini (✉)
Centro Studi G.B. Carducci, Corso Marconi 37, 63900 Fermo (FM), Italy
e-mail: centrostudicarducci@gmail.com

G. Minati et al. (eds.), *Towards a Post-Bertalanffy Systemics*, Contemporary Systems Thinking, DOI 10.1007/978-3-319-24391-7_8

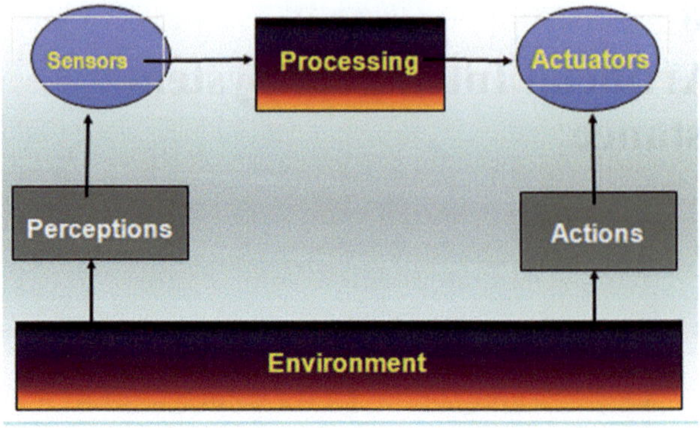

Fig. 8.1: Intelligent system

AIS, named EPAS (Elderly People Assistant System), conceived to support the elderly in their needs [1, 2], like mobility, memory, leisure and health.

Research in the field of intelligent systems is moving towards the creation of humanoids. These will be able to walk on uneven ground in the real world, open close doors with great autonomy, do not lose mobility when they fall.

The Industrialization of humanoid robots is the target in the near future. Meanwhile, there is a tendency towards Intelligent Systems Reliable for everyday life. It is to develop technology-based systems with high reliability, high security functional, with assessment and risk management, able to overcome human error, evolved in terms of 'Physical Human-Robot Interaction'.

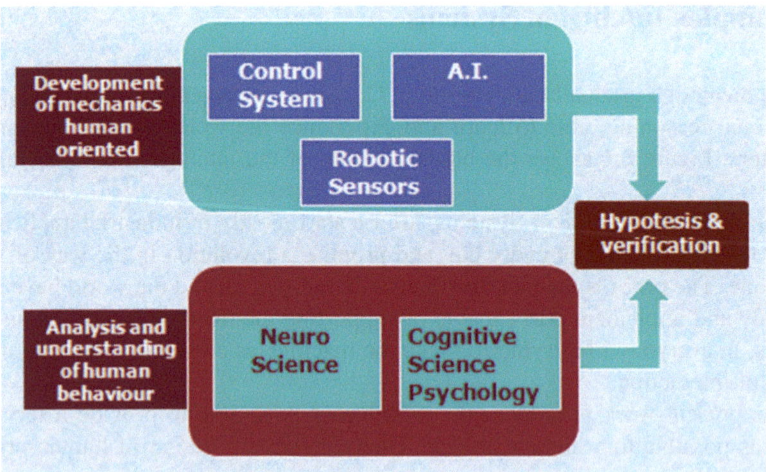

Fig. 8.2: Human oriented AIS

3 EPAS

Build a robot that present human-like behavior is terribly complicated, but fascinating. Being obviously impossible to build a robot that shows human feelings and emotions, we can still make the robot that behaves 'apparently' like a human, thus improving our relationship with the machine. We can create robots that read, that recognize objects and interact with the surrounding environment. In designing AIS that interacts in an almost human manner with older people, we have to use, at least in part, the technology of the humanoids. It must be able to interact with humans in a natural way: speak, answer, recognize people, and use human behavior, such as emotional. It must also act as assistant to people who, because of their advanced age, have problems with memory, reading and sometimes mobility. For this last reason, it must be, rather than a biped, an AIS which smoothly moves on wheels, capable of transporting persons and easy to drive, with manual and oral controls. In addition, the cognitive aspects are very important.

To address the problems of appearance and of behavior, two approaches are necessary: the Robotic one and the Cognitive one. The Robotic seeks to build a robot very close to man, based on the Cognitive Science. While the Cognitive Science uses the robot to test hypotheses about human behavior. For both approaches applies the block diagram of Fig. 8.2 which highlights the complexity of the system.

A fundamental property of the android's science is the existence of the uncanny valley. In Fig. 8.3 it is represented the hypothesis of the "Uncanny Valley". This says that the degree of confidence between man and robot increases with the appearance and behavior of the human type. But at a certain level of these, the degree of confidence falls sharply.

For example a "zombie" is close to the uncanny valley, as well as the child android, that is often achieved by making a copy of an existing child. The research in this field attempts to verify the existence of the uncanny valley and explore ways of overcoming the problem. Clearly EPAS is far from this valley. It needs to have a pleasant interaction with the elder, so that the machine is well accepted for a large

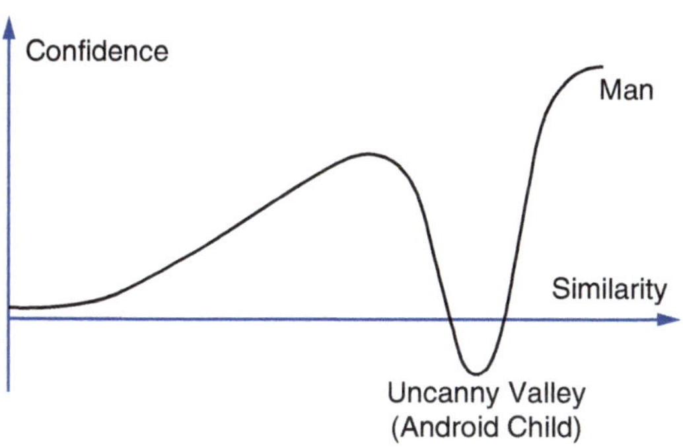

Fig. 8.3: Confidence versus similarity in AIS

proportion of services, done by it, instead of a "human assistant". The control software of EPAS, called MIND (Monitoring of Intelligence and Demand) (Fig. 8.4) controls the speaking, reading, mathematics, vision, colors, sound, automation and sensors. The control is based not only on manual controls, even on the recognition of sentences spoken, on read the facial expressions and on recognition of 3-D objects. MIND is conceived to interact with the surrounding environment, to process and record the information in its internal memory. EPAS is AIS for elderly people in an indoor environment, in a first version, and in an outdoor environment in a second version. The overall aim of EPAS is to allow independence and autonomy in everyday life for the elderly and disabled. It is designed equipped with man-machine interface oriented to: (1) make it easier and pleasant user interaction with the machine, (2) provide a range of transport services, entertainment, intelligent support to the user who can ambulate or to one that cannot ambulate. It must meet the criteria of "Assistive Technology", that allow maximum independence to access commands and monitors. Capable of responding to commands: vocal, gestural, or typed [4]. It has sensors able to feel temperature, pressure, etc. and, in general, programmable "alerts". EPAS is able to detect obstacles in the path; signaling and revealing the fixed and mobile. Launches also sound "alerts". It is conceived in two versions: one with two wheels and one with four wheels. Both versions use platforms such segway. You can be on board and control the movements with natural movements of the body, using Gyroscopes System platforms. In Fig. 8.4 it is shown the diagram of the MIND [3, 5–10] control software. EPAS has two cameras and a set of microphones. The cameras with intelligent software of vision [11–17] allow seeing the environment and obtaining a map of this. The microphones allow you to localize the sources and then the speaker. The AIS locates sound source, activates the Intelligent Vision System, recognizes the face and tries to approach the user.

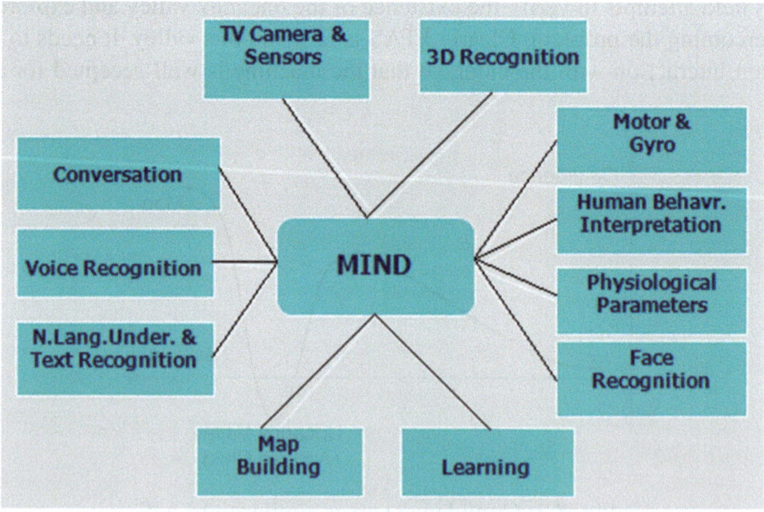

Fig. 8.4: MIND software

Fig. 8.5: EPAS

EPAS is designed to use a platform of Segway type (Fig. 8.5) and able to move around outdoors, overcoming simple architectural barriers. It is endowed of a number of security systems which avoid falls, collisions and hazards. Besides MIND can puts into action a set of strategies and if necessary can stop moving and launch acoustic "alerts" or alarms through a wireless network. Finally EPAS is conceived to be trained, for learning to cooperate [18–20] with other similar AIS, or with humans.Among its equipments, are included entertainment facilities, like playing, listening radio programs, watching TV programs, use of multimedia and Internet.

References

1. Lacey, G., & Dawson-Howe, K. M. (1998). The application of robotics to a mobility aid for the elderly blind. *Robotics and Autonomous Systems*, (p. 23).
2. US Department of Health and Human Services. (1999). *Health and aging chart book*, Health.
3. Tascini, G., Montesanto, A., Penna, M. P., & Pessa, E. (1999). A hybrid architecture for autonomous robots undertaking goal-directed navigation. In *Atti del workshop apprendimento e percezione nei sistemi robotici*, Parma.
4. Lakemeyer, G. (Ed.). (2000). *Notes Second International Workshop on Cognitive Robotics*, Berlin.
5. Singh, S., Kearns, M., Litman, D., & Walker, M. (2000). Reinforcement learning for spoken dialogue systems. In *NIPS, 2000*.
6. Tascini, G., Monteanto, A., & Palombo, R. (2000). Video description by automatic context extraction. In *IEEE Proceedings of International Symposium on Visual Languages* (pp. 89–ff). Los Alamitos, CA: IEEE Press.
7. Tascini, G., Regini, L., Montesanto, A., & Puliti, P. (2000). Interazione sistema autonomo-mondo in ambiente virtuale. In *Atti del settimo convegno dell'associazione Italiana per l'Intelligenza artificiale*, Milano (pp. 119–122).

8. Poupart, P., Ortiz, L. E., & Boutilier, C. (2001). Value-directed sampling methods for monitoring POMDPs. In *Proceedings of the Seventeenth Annual Conference on Uncertainty in Artificial Intelligence (UAI-01)*, Seattle (pp. 453–461).
9. Tascini, G., Montesanto, A., Palombo, R., & Puliti, P. (2001). Behind the image sequence: The semantics of moving shapes. In *Lecture notes in computer science* (Vol. 2059, pp. 619–629). Berlin: Springer.
10. Tascini, G., Montesanto, A., & Puliti, P. (2001). Reactive navigation using reinforcement learning techniques in situations of POMDPs. In *Lecture notes in computer science* (Vol. 2085, pp. 444–450). Berlin: Springer.
11. Tascini, G., Montesanto, A., Puliti, P., & Rabascini, N. (2002). Reactive navigation based on self-organised visual information. In G. Minati & E. Pessa (Eds.), *Emergence in complex cognitive, social and biological systems* (pp. 193–204). New York: Kluwer Academic/Plenum Publishers.
12. McCarthy, C.E., & Pollack, M. (2002). A plan-based personalized cognitive orthotic. In *AIPS-2002*.
13. Tascini, G., De Cesaris, I., & Montesanto, A. (2002). Emergence of paths in robot simulation. MITOS (Eds.), *Proceedings of 5th System Science Congress*, Heraklion, 2002 (pp. 59.1-59.1–59.11).
14. Baldassarri, P., Puliti, P., Montesanto, A., & Tascini, G. (2003). Self-organizing maps versus growing neural gas in robotic application. In *Lecture notes in computer science* (Vol. 2686). Berlin: Springer.
15. Tascini, G. (2004). Virtual navigation. In *Imaging science, systems and technology* (pp. 548–554). La Vegas: CSREA Press.
16. Tascini, G., Baldassarri, P., Montesanto, A., & Puliti, P. (2004). Robot localization using incremental neural networks. In *Convegno AI*IA04*, Perugia, 2004.
17. Montesanto, A., Tascini, G., Puliti, P., & Baldassarri, P. (2006). Navigation with memory in a partially observable environment. *Robotics and Autonomous Systems, 54*(1), 84–94.
18. Tascini, G., & Montesanto, A. (2006). Emergence of cooperation-competition between two robots. In G. Minati, E. Pessa, & M. Abram (Eds.), *Systemics of emergence: Research and development* (pp. 317–340). New York: Springer.
19. Tascini, G. (2008). Meta genetic behaviour emergence in evolutionary simulation. In *Proceeding of Grand Challenges in Computational Biology*. Joint BSC - IRB workshop, Barcelona, June 2008.
20. Tascini, G. (2012). Selfish gene or altruistic organisms? In G. Minati, M. Abram, & E. Pessa (Eds.), *Methods, models, simulations and approaches towards a general theory of change* (pp. 187–202). Singapore: World Scientific.

Chapter 9
From Systemic Complexity to Systemic Simplicity: A New Networking Node Approach

Rodolfo A. Fiorini

1 Introduction

Contemporary human made applications and systems can be quite fragile to unexpected perturbation because Statistics can fool you. Our computational information contemporary classic systemic and instrumentation tools (developed under the positivist reductionist paradigm, for short "Science 1.0") are totally unable to capture and to tell the difference between an information-rich message (optimally encoded message) and a random jumble of signs that we call "noise": in scientific literature this distressing dilemma is called "Computational Information Double-Bind" [4]. In fact, classical experimental observation process, even in highly ideal operative controlled condition, like the one achieved in contemporary most sophisticated and advanced experimental laboratories like CERN [1] can capture just a small fraction only of overall ideally available information, from unique experiment. The remaining part is lost and inevitably "dispersed" through environment into something we call "background noise" or "random noise" usually, in any scientific experimental endeavor [3]. The amount of information an individual or a system can acquire in an instant or in a lifetime is finite, and minuscule compared with what the milieu presents; many questions are too complex to describe, let alone solve, in a practicable length of time. Usually, information uncertainty and incompleteness are approached by classic probability risk management techniques. Unfortunately, unpredictable changes cannot be managed by contemporary risk management techniques and they can be very disorienting at enterprise level [15]. These major changes, usually discontinuities referred to as fractures in the environment rather than trends, will largely determine the long-term future of organization. They need to be handled, as opportunities, as positively as possible. We need

R.A. Fiorini (✉)
Department of Electronics, Information and Bioengineering (DEIB),
Politecnico di Milano University, 32, Piazza Leonardo da Vinci, 20133 Milano, Italy
e-mail: rodolfo.fiorini@polimi.it

G. Minati et al. (eds.), *Towards a Post-Bertalanffy Systemics*, Contemporary Systems Thinking, DOI 10.1007/978-3-319-24391-7_9

more robust, resilient and antifragile [14] application to be ready for next generation systems. In turn, they are mandatory to develop antifragile self-organizing and self-regulating system further. In full agreement with Taleb, our basic idea is that an assessment of system fragility (and control of such fragility) is more useful, and more reliable, than probability risk management and data-based methods of risk detection and prediction. Main attention focus should not be to attempt to predict black swan events, but to build system robustness against negative ones that occur and be able to exploit positive ones. In the past five decades, trend in Systems Theory, in specialized research areas, has slowly shifted from "General System Theory", introduced by Ludwig von Bertalanffy and classic single domain information channel transfer function approach ("Shannon's channel", introduced in 1941, associated to traditional computational model under either additive or multiplicative perturbation hypothesis), to the more structured "ODR Functional Sub-domain Transfer Function Approach" (by Observation, Description and Representation Functional Blocks) [7]. Nevertheless, if careful information conservation countermeasure is not provided at each step, from source to destination, ODR active transmission channel could suffer from the same problem, discussed earlier.

2 Achieving Systemic Resilience and Antifragility

According to fresh Computational Information Conservation Theory (*CICT*) result, scientific community has acquired new awareness about traditional rational number system $\boldsymbol{Q}$ numeric properties, quite recently [5, 6]. Thanks to this line of generative thinking, it is possible to realize that traditional rational number system can be even regarded as a highly sophisticated open logic, powerful and flexible LTR ("Left-To-Right") and RTL ("Right-To-Left") formal language of languages, with self-defining consistent words and rules, starting from elementary generators and relations [5, 6]. ODR approach can take advantage immediately from those properties to develop system computational functional closures to achieve information conservation countermeasure at each operative step automatically. Then, all computational information usually lost by classic information approach, based on the traditional noise-affected data stochastic model only, can be captured and fully recovered to arbitrary precision by a corresponding complementary codomain, step-by-step. Theoretically, codomain information can be used to correct any computed result, achieving computational information conservation (virtually noise-free data), according to *CICT* Infocentric World-view [4]. In this way, system resilience and antifragility can be developed quite easily [4]. From an application realization point of view, Canadian ecologist Crawford Stanley (Buzz) Holling introduced important ideas in the application of ecology and evolution, including resilience, adaptive management, the adaptive cycle, and panarchy [10, 11]. Following Holling's approach, our main idea is to introduce a new networking node able to bind known information to the unknown one in coherent way. Then, unknown "environmental noise" or/and local "signal input" information can be aggregated to known "system internal control status" information, to provide self-structuring synthetic attractor point and functional closure. In this way, a self-organizing landscape of self structuring attractor points can be obtained, which a balanced "Operating Point" can always emerge from, as a new Trans-disciplinary Reality Level, from the interaction of an ideal asymptotic dichotomy: two coupled complementary irreducible information computational management subsystems. Due to its intrinsic self-scaling properties, this system approach can be applied at any system scale: from single quantum system application development to full system governance strategic assessment policies and beyond [8].

3 Articulated Interaction by Recursive Information Aggregation

In a continuously changing operational environment, even if operational parameters cannot be closely predefined at system design level, we need to be able to design reliable self-organizing, self-regulating and self-adapting system quite easily anyway. We already know about "self-reference" in mathematics as a statement that refers to itself, for example, as a set that contains itself. Traditionally, such statements lead to paradox, a form of inconsistency. In the informal fallacies self-referential statements are considered poor form. That is true in mathematics and arithmetics when you use a continuous support approach and do not take advantage from the finiteness limitations of your real computational resources [4]. In short, it is necessary to find a mathematical method to aggregate the external information coming from environment with system internal information, in an efficient and fast way that is both immune from computational polynomial mirroring and classic noise effects, by design. Furthermore, the mathematical method would possess anticipatory properties needed to the realization of system able to interact with its environment in real time (leading property). A mathematical method that meets these requirements (articulated information aggregation with system information anticipation and no computational polynomial mirroring and noise effects) is a simple recursive relation, quite well known in mathematics, since long time!

3.1 The Root of Systemic Networking Articulation

In his book "Liber abaci", for the first time Fibonacci (1170–1250) introduced the concept of recursive sequence to the Western culture, with the famous sequence:

$$0, 1, 1, 2, 3, 5, 8, 13, 21, 34, 55, 89, 144, \ldots \tag{9.1}$$

in which each term is the sum of the two preceding ones and the numerical sequence composition law can be written:

$$F_n = F_{n-1} + F_{n-2}, with\ \ F_0 = 0\ \ and\ \ F_1 = 1, \tag{9.2}$$

and in general, $F_n = k_1 * F_{n-1} + k_2 * F_{n-2}$, where $k_1, k_2 = 0, 1, 2, \ldots \infty$, $k_1, k_2 \in \boldsymbol{N}$, for Generalized Fibonacci Sequences. For original Fibonacci sequence $k_1, k_2 = 1$.

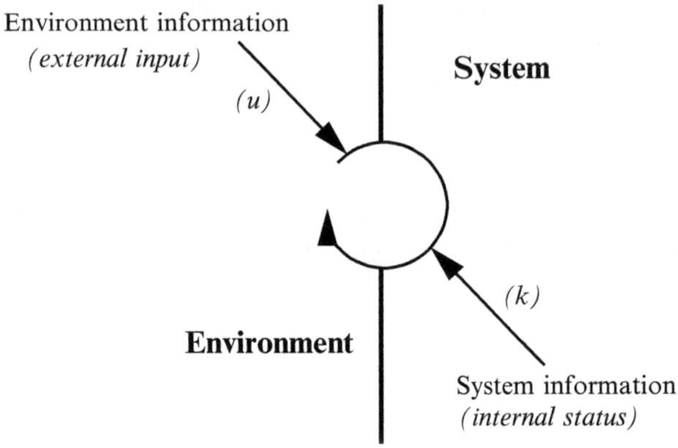

Fig. 9.1: System interaction with environment by external input (u) and internal system status information (k) by recursive aggregation [2]

As a starting point, this relation can be thought as the aggregation of external information $[u_1, u_2]$ to internal system status information $[k_1, k_2]$. Therefore, recursive sequence information aggregation offers at least three operational advantages over usual direct polynomial quantification. First, recursive sequence information aggregation does not suffer from the computational polynomial mirroring effect. Second, its asymptotic convergence properties of the ratio of successive terms allows the creation of a structured system behavioral space similar, in computational behavior, to living organism homeostasis (i.e. the automatic selection of environment minimum perturbation level that allows optimal interaction between external information from environment (u) and system internal status information (k), as evidenced by Holling [11]. We name this new networking node "Recursive Interactor" (Fig. 9.1). Third, recursive sequence information aggregation represents a computational method with intrinsic computational anticipatory properties, because it is possible to structure anticipatory computation for successive sequence terms arbitrarily. Then taking any positional index it is possible to compute not only the next one but also those ones at any distance from the current position in an anticipatory way. Moreover, the ratio of Fibonacci sequence two consecutive terms, according to the relation:

$$\lim_{n \to \infty} \frac{F_{(n+1)}}{F_{(n)}} = \Phi, \tag{9.3}$$

asymptotically converges to the golden number, property discovered by Kepler, Tattersall [16] providing us with recurrence relation functional closure. For this simple case, the recursive aggregation law of information can be easily extended to three consecutive terms (starting by trinomial $0,0,1$, third order relation, $m = 3$), four successive terms (starting by quadrinomial $0,0,0,1$, fourth order relation, $m = 4$), j successive terms (starting by m-nomial formed by j zeroes, plus one 1, $(j+1)$th order relation, $m = j+1$), and in general we can write, in compact form, recursive relations as function of three parameters, $a_n = a(m,k,u)$, as specified previously, where m is the recurrence relation order. In this way, we can allow for articulated information aggregation in a networking environment and computing their related asymptotic functional closures immediately. Therefore, it is possible to compute recursively further numerical sequences asymptotically converging to irrational limits or completely diverging as function of external information input. In this way system can search automatically for a minimum environmental perturbation level (system internal status) useful to insure sequence asymptotically convergence to get vital information from system environment (self-regulation and learning as quest for the difference that makes the difference, probing by probing ...). Furthermore, system can even automatically self-organize and structure numeric families with different numerical closure to conserve overall system information (Generalized Fibonacci Systems, and information conservation by irreducible complementary system) [4]. But, never forget one of Robert Rosen's fundamental lessons: human formal systems are unable to capture enough knowledge to model natural system completely [13]. You have to model natural system with deep awareness to grasp a part of it only! What you will always get is finite precision uncertainty awareness from never-ending natural evolution representation. So the best you can do is to find your best strategic management solution to handle arbitrary incomplete knowledge and finite precision system uncertainty [4].

3.2 System Output Information Anticipation

Recursive sequence represents a convenient mathematical method that holds anticipatory proprieties. In fact, it is possible to implement the anticipatory computation of any recursive sequence's term. Taking arbitrarily any current positional index, we can compute not only the next term but also terms at a certain distance from the current one in an anticipatory way, compared with the current positional index, by implementing its primary relation recursion conveniently. Specifically, starting from the recursive rule that indicates the next term to the current one, it is possible to structure a set of rules that allows to obtain recursive sequence terms at different distance, defining a set of registers that, working in parallel, are able to provide values with the desired anticipation level immediately. For example, considering second order ($m = 2$) recursive relations $a_n = a(2,k,u)$, where k and u are 2-d vectors $[k_1, k_2]$ and $[u_1, u_2]$ respectively, as depicted in Fig. 9.1, and where the $(n+1)$th to the current nth term is obtained in the following way, we have:

$$a_{n+1} = k_1 * a_n + k_2 * a_{n-1}, \tag{9.4}$$

where $a_{n-1} = u_1 = 0$ and $a_n = u_2 = 1$, $k_1 = k_2 = 1$, for the Fibonacci sequence. Then, we can specify recursion derived relations to compute appropriate terms at any arbitrary distance from the current position n. As an example, we define the following relations that are valid for computing terms to the distance $n+5$ from the current one, depending, for instance, on parameters $[k_1, k_2] = [1, 1]$ and $[k_1, k_2] = [2, 2]$. These recursive relations can be used in parallel, respectively, to provide the terms of sequence in an anticipatory way simultaneously. So in the case of $[k_1, k_2] = [1, 1]$, aggregation rules are as follows:

$$\begin{aligned} a_{n+2} &= 2 * a_n + 1 * a_{n-1} \\ a_{n+3} &= 3 * a_n + 2 * a_{n-1} \\ a_{n+4} &= 5 * a_n + 3 * a_{n-1} \\ a_{n+5} &= 8 * a_n + 5 * a_{n-1} \\ &\ldots \end{aligned} \tag{9.5}$$

In the case of values $[k_1, k_2] = [2, 2]$ we obtain the following aggregation rules:

$$\begin{aligned} a_{n+2} &= 6 * a_n + 4 * a_{n-1} \\ a_{n+3} &= 16 * a_n + 12 * a_{n-1} \\ a_{n+4} &= 44 * a_n + 32 * a_{n-1} \\ a_{n+5} &= 120 * a_n + 88 * a_{n-1} \\ &\ldots \end{aligned} \tag{9.6}$$

In general for any vector $[k_1, k_2]$ and for any order m, it is always possible to formulate rules associated to their primary recursion relation: these rules allow the parallel anticipatory computation of recursive sequence terms at any distance from the current position term n.

4 Open Logic and Closed Logic Subsystem

Irrational numeric limit families, identified by converging recursive numeric sequences allow the structuring of a mathematical Baire's Space to manage numeric information useful to synthesize quick and raw system primary response "to survive" (Apprehension, Open Logic Section, see Fig. 9.2). To synthesize more organized and articulated, but slower, system response "to learn", it is necessary to structure recursive information into an "ordered polynomial sequence", by "polynomial weighing". Polynomial weighing is a mapping to achieve "coherent perception" (Organization, Closed Logic Section, see Fig. 9.2) [2].

Polynomial weighing is key mapping operation to map recursive sequence information representation into polynomial format corresponding to combinatorially optimized exponential cyclic sequence (*OECS*), folded into rational number operative representation [4]. So, we get a sequence of different structuring operations

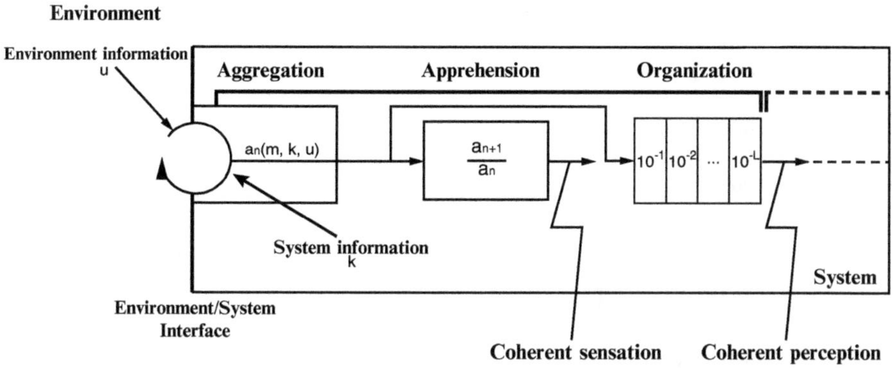

Fig. 9.2: Interaction interface system (IIS) reference architecture [2]

to get external information more and more coherently formatted to system internal status to arrive to a system "coherent perception" of external information. We get the overall *IIS* (Interaction Interface System) reference architecture as depicted in Fig. 9.2.

5 *VEDA* Simulation Environment

The simple *IIS* by recursive information aggregation method of Fig. 9.2 can be proposed for advanced *ISS* (Inner Safety System) in biomedical and advanced healthcare development system modeling, from neuroperception, bio-transduction functions, to organ level modeling and beyond. To design, analyze and test *IIS* and *ISS* system properties, a simulation environment has been designed, developed and implemented, programmed in MATLAB language, called *VEDA* (Visualization of Evolutionary Dynamics Application), at Politecnico di Milano University. *VEDA* system dynamics simulation toolbox offer a high level simulation flexibility by user-optimized graphic interface to get easier simulation task, to design, analyze and synthesize complex dynamical system behavior.

In this way, it is possible to study natural complex dynamics simulation, to verify and validate through numerical computation and displaying the behavior of all subsystems that compose the final combined overall system performance. So, according to desired system parameters, to validate the choice of optimal parameters set, for a specific embodiment, is quite straightforward [2]. As a simple example, let us examine a fifth order system dynamics with a divergent master dynamics (LTR, upper right side of Fig. 9.3, labeled **mr**). You see, it may recall a simple exponential response, resembling closely a P100 onset recording, from usual EEG and ERP preprocessing, from a normal subject under a cognitive task. Expliciting the associated RTL sequence, on the left side of Fig. 9.3 (**ml**), you get a point sequence more difficult to grasp immediately than the previous one. *VEDA* simulation

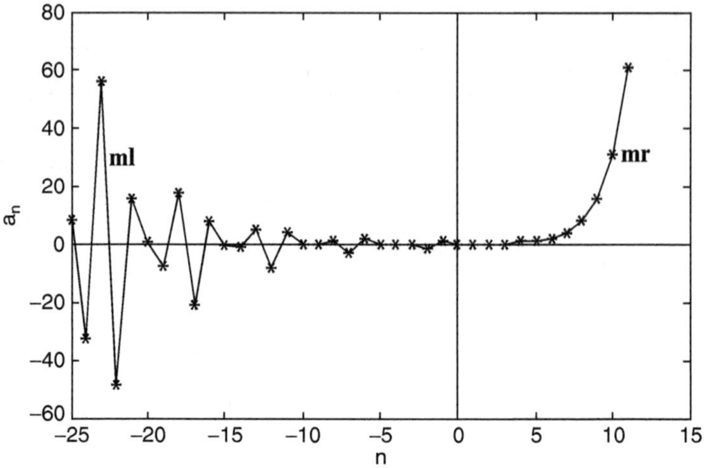

Fig. 9.3: Divergent **LTR** master dynamics (**mr**, *upper right side*) generated by apparently irregularly arranged point sequence (**ml**, on the *left side*)

toolbox immediately shows that those, apparently irregularly arranged, points can be structured by the coherent combination of five slave subsequences, as reported in Fig. 9.4 (left slave subsequences, **a**–**e**). Furthermore, by LTR to RTL reflection of the five left slave subsequences with respect to vertical axis, it is possible to grasp the

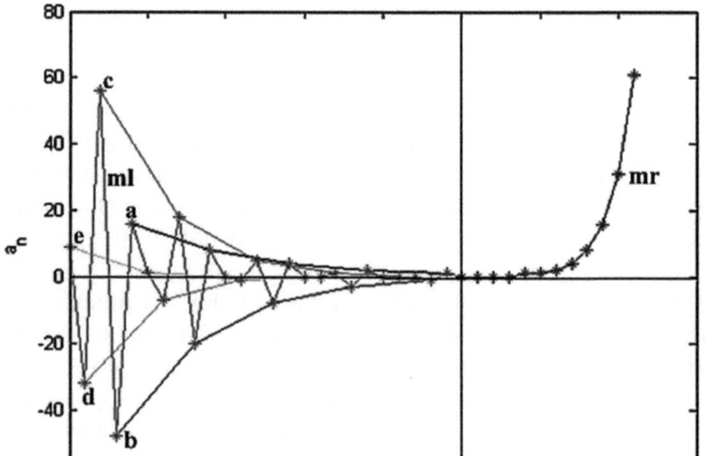

Fig. 9.4: Coherent combination of five **RTL** slave subsequences (**a**–**e**) to synthesize the unique master sequence (**mr**) on the *right side* (see text)

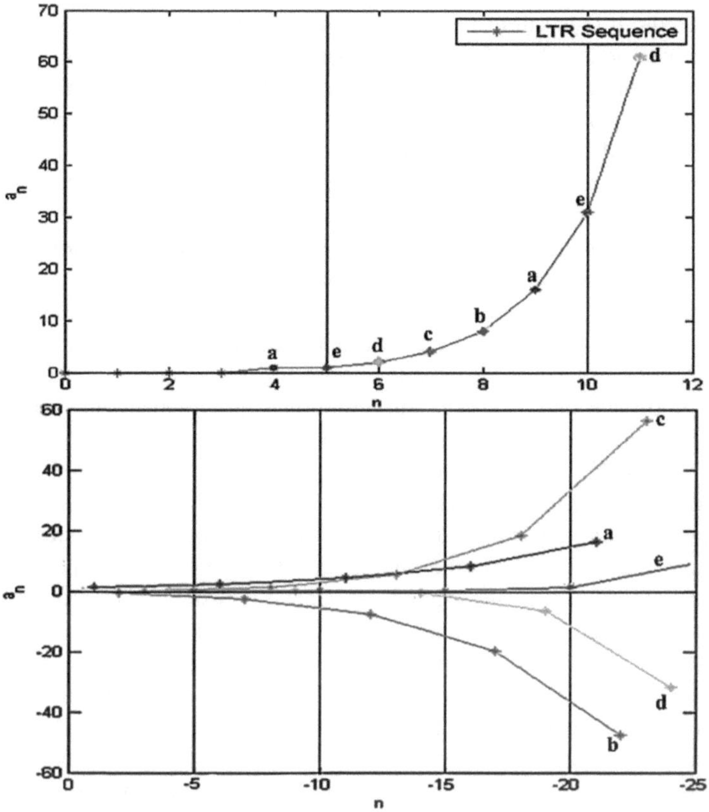

Fig. 9.5: The comparison of the **LTR** to **RTL** reflection of the left slave subsequences (**a**–**e**) with respect to vertical axis (*lower diagram*) to the unique coherent composite right master sequence (**mr**, *upper diagram*) allows for grasping the dynamics of the sequential coherent combination of the five left slave subsequences (see text)

dynamics of the sequential coherent combination of the five slave subsequences into the unique coherent final composite right master sequence, as offered by Fig. 9.5.

6 Summary and Conclusion

The major added value of our approach is provided by our new idea of system articulated interaction, defined as inner and outer system information aggregation. It can allow both quick and raw system response (to survive and grow) and slow and accurate information unfolding for future response strategic organization (to learn and prosper) by coherently formatted operating point [2]. Now, according to previous discussion, it is possible, at systemic level, to envisage a post-Bertalanffy Systemics Framework, with multiscaling properties, able to deal with problems of different scale complexity in a generalized way when inter-disciplinarity consists, for instance, of a disciplinary reformulation of problems, like from biological to chemical, from clinical research to healthcare, etc., and trans-disciplinarity is related to the study of such reformulations and their properties. As shown from Fig. 9.6, our innovative system interaction modality, called "Recursive Interactor", corresponds to fourth order biomedical cybernetics. Now, new advanced systemic information application can successfully and reliably manage a higher system complexity than contemporary ones, with a minimum of design constraints specification and less system final operative environment knowledge at design level. For instance, according to Fig. 9.2, at brain level, it is possible to refer to "Le-Doux circuit" (Logical Aperture) for emotional behavior (i.e. fear, emotional intelligence, etc.) and to "Papez circuit" (Logical Closure) for structured behavior (i.e. rational thinking, knowledge extraction, etc.) [12] EI (Emotional Intelligence) and EC (Emotional Creativity) [9] coexist at the same time with RT (Rational Thinking) in human mind, sharing the same input environment information. Then, operating point can emerge as a trans-disciplinary reality level from the interaction of two coupled complementary irreducible, asymptotic ideal subsystems. To behave realistically, overall system must guarantee both Logical Aperture (to get EI and EC, to survive and grow) and Logical Closure (to get RT, to learn and prosper), both fed by environmental "noise" (much better, from what human beings call "noise") [4]. In fact, natural living organism does perturb its environment, but only up to the level it is perturbed in turn by its own environment both to survive and grow, no more [11]. Due to its intrinsic self-scaling properties, this system approach can be applied at any system scale: from single quantum system application development to full system governance strategic assessment policies and beyond [7]. We can use the same nonlinear logic approach to guess a convenient basic architecture for *ALS* (Anticipatory Learning System) to get realistic modeling of natural behavior to be used in High Reliable Organization (HRO) application development [8]. *VEDA* is an original simulation environment developed to validate complex dynamical system model. *ALS* and *ISS* are two pivotal concepts to develop safer, more resilient, effective and efficient solutions for competitive safety systems and human wellbeing. As a matter of fact their basic operational concepts can be conveniently and successfully extended to many other advanced Business and HRO application areas, with no performance or economic penalty, to develop more and more competitive application. Expected impacts are multifarious and quite articulated at different system scale levels: at theoretic level, major one is that, for the first time, Biomedical Engineering ad Bioengineering ideal system categorization levels can be matched exactly to practical system modeling interaction styles (Fig. 9.6), with no paradigmatic operational ambiguity and information loss, even for living organism application. The present paper is a relevant contribute towards a new General Theory of Systems to show how homeostatic equilibria can emerge out of a self-organizing landscape of self-structuring attractor points.

BIOMEDICAL CYBERNETIC ORDER	INTERACTION STYLE	GRAPHIC SYMBOL
Zero	Pure Spectator	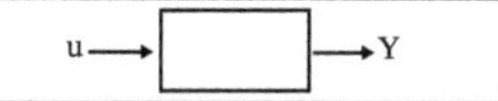
First	Ergodic Observer	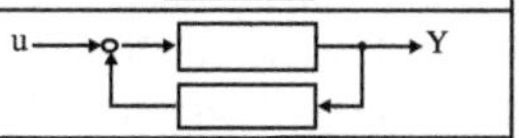
Second	Pulsed Egocentric Interactor	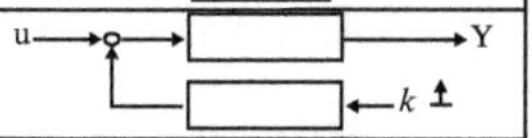
Third	Iterated Egocentrice Interactor	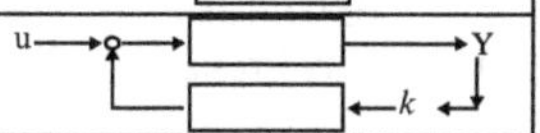
Fourth	Recursive Interactor	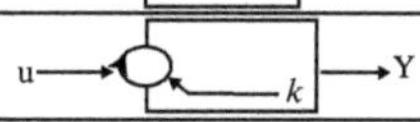

Fig. 9.6: Our final post-Bertalanffy systemics framework [7]

Acknowledgements We like to thank our friends (in alphabetical order) Fortunato Tito Arecchi, Renato Betti, Claudio Citrini, Piero De Giacomo, Emilio Del Giudice, Paolo Dulio, Luciano L'Abate, Joseph E. LeDoux, Ignazio Licata, Pierluigi Marconi, J. David Powell, Beverly Rubik, Adolph Smith and Giuseppe Vitiello for useful discussions. We want to dedicate this work to the memory of Emilio del Giudice. So much we regret his recent, sudden and premature delocalization to cosmic vacuum.

References

1. C.E.U. European Laboratory for Particle Physics, Conseil Européen pour la Recherche Nucléaire (2013). Accessed 30 Jan 2015. http://home.web.cern.ch/about/updates/2013/11/atlas-sees-higgs-boson-decay-fermions
2. Collini, A. V., & Cesario, C. (2012). *Sistemi a sicurezza intrinseca per la sanitá*. Master of Science thesis, Biomedical Engineering Department, Politecnico di Milano, Milano.
3. Fiorini, R. A. (1994). *Strumentazione biomedica: sistemi di supporto attivo*. Milano: CUSL, Collana Scientifica.
4. Fiorini, R. A. (2014). How random is your tomographic noise? A number theoretic transform (NTT) approach. *Fundamenta Informaticae, 135*, 135–170.
5. Fiorini, R. A., Condorelli, A., & Laguteta, G. (2013). Discrete tomography data footprint reduction via natural compression. *Fundamenta Informaticae, 125*, 273–284.
6. Fiorini, R. A., & Laguteta, G. (2013). Discrete tomography data footprint reduction by information conservation. *Fundamenta Informaticae, 125*, 261–272.
7. Fiorini, R. A., & Santacroce, G. F. (2014). A post-Bertalanffy systemics healthcare competitive framework proposal. In *Proceedings of 12th International Conference on Informatics, Management, and Technology in Healthcare*, Athens, July 10–13.
8. Fiorini, R. A., & Santacroce, G. F. (2014). Safety and effectiveness health systemic governance by HICT natural framework. In *Proceedings of International Conference on Biomedical Informatics*, Milano, May 8–9.
9. Goleman, D. J. (1995). *Emotional intelligence: Why it can matter more than IQ*. New York: Bantam Books.
10. Gotts, N. M. (2007). Resilience, panarchy, and world-systems analysis. *Ecology and Society, 12*(1), 24. Accessed 30 July 2014. http://www.ecologyandsociety.org/vol12/iss1/art24/

11. Gunderson, L. H., & Holling, C. S. (Eds.). (2002). *Panarchy: Understanding transformations in human and natural systems*. Washington, DC: Island Press.
12. Ledoux, J. (2002). *Synaptic self, how our brains become who we are*. New York: Viking Penguin.
13. Rosen, R. (1985). *Anticipatory systems*. Oxford: Pergamon Press.
14. Taleb, N. N., & Douady, R. (2013). Mathematical definition, mapping, and detection of (anti) fragility. *Quantitative Finance, 13*(11), 1677–1689.
15. Taleb, N. N., & Goldstein, D. G. (2012). The problem is beyond psychology: The real world is more random than regression analyses. *International Journal of Forecasting, 28*(3), 715–716.
16. Tattersall, J. J. (2005). *Elementary number theory in nine chapters* (2nd ed., p. 28). Cambridge: Cambridge University Press.

Chapter 10
Formal Concept Analysis in Statistical Hypothesis Testing

Eraldo Nicotra and Andrea Spoto

1 Introduction

Hypothesis Testing consists of solving some kind of statistical test regarding the equality of group-referred parameters. Usually, this kind of procedure can be subdivided in three main steps: hypothesis definition, application of algebraic transformations over sampled parameters and conclusion. Solvers may present a wide variety of solution's patterns related to students' ability to solve the entire problem. The relation between the response patterns and the capability of each solver to master each part of the exercise could be mapped through a partial order and conveniently represented using the theoretical framework of Formal Concept Analysis (FCA) [1, 2] whose main concepts are introduced below.

The first basic notion of FCA is the *formal context* defined as a triple (G,M,I) where G is a set of *objects*, M is a set of *attributes* and I is a *binary relation* between the set of objects and the set of attributes. A formal context is usually represented by a Boolean matrix where each row is an object and each column is an attribute. Whenever a 1 is present in the entry (g,m), it means that the relation gIm for the specific g and m holds. Between objects and attributes of a formal context a *Galois connection* is defined. For all the sets $A \subseteq G$ and $B \subseteq M$, the following two transformations define the Galois connection:

$$A' := \{m \in M | gIm, \forall g \in A\} \tag{10.1}$$

$$B' := \{g \in G | gIm, \forall m \in B\} \tag{10.2}$$

E. Nicotra (✉)
Dipartimento di Pedagogia, Psicologia, Filosofia, University of Cagliari, Via Is Mirrionis 1, Cagliari, Italy
e-mail: enicotra@unica.it

A. Spoto
Department of General Psychology, University of Padua, Via Venezia 8, Padua, Italy
e-mail: andrea.spoto@unipd.it

G. Minati et al. (eds.), *Towards a Post-Bertalanffy Systemics*, Contemporary Systems Thinking, DOI 10.1007/978-3-319-24391-7_10

In words, A' is the collection of all the attributes that all the objects in A have in common. Dually B' is the collection of all the objects that possess all the attributes in B. It is now possible to introduce the fundamental notion of *formal concept*, that is a pair (A,B) that satisfies the following two conditions: $A = B'$ and $B = A'$. The *extent* A of the formal concept contains exactly those objects of G that have all the attributes in B; the *intent* B of the formal concept includes exactly those attributes satisfied by all the objects in A. A sub-concept super-concept relation is then defined in the following way:

$$(A_1,B_1) \leq (A_2,B_2) \rightarrow A_1 \subseteq A_2 \tag{10.3}$$

or equivalently:

$$(A_1,B_1) \leq (A_2,B_2) \rightarrow B_1 \supseteq B_2 \tag{10.4}$$

In words, a concept is of a lower level when it has a larger extent (or equivalently a smaller intent). The concepts of a context form a *complete lattice* [3] that is called the *concept lattice* of (G,M,I). The intents of a concept lattice are closed under intersection i.e. each intersection of sets of attributes is included in the lattice. In the present application the set of objects of our context consists of the response patterns to an exercise of "Hypothesis Testing for Paired Samples t-test", while the set of attributes consists of the subcomponents of the exercise that each pattern involves. In the Methods section we will go into more details about how the exercise has been divided in its subcomponents.

So far we introduced the deterministic part of the theoretical framework conceived for the present work. In evaluating knowledge some sort of probabilistic model has to be used in order to account for the variability of observed data. We decided to refer to the general framework of Item Response Theory and, more specifically, to the Rash Model [4].

The use of the *simple logistic model* was functional to obtain a probabilistic evaluation of the occurrence of each single response pattern observed to the task administered to the students. The application of the Rash simple logistic model, for dichotomous items allowed us to map each single deterministically detected items' structure obtained from the FCA solution into a probabilistic measurement system. This system, more realistically, conveys information about the probability to observe each specific response pattern, as the number of observations becomes sufficiently large. In details, the simple logistic model plays a fundamental role in discovering the measurement level of two parameters involved in items solution: the ability of a person to solve a problem and the difficulty of the asked items. We can now introduce some formalism to expose the mathematical structure of the Rasch's simple logistic model. The basic assumption of the model imposes that the probability to obtain a right solution of the proposed problem, for each respondent, can be related to the difficulty of the item and to the ability of the respondent. The more an individual is able the higher is the probability to obtain the correct solution to the problem, given the specific item difficulty. The mathematical relationship involved in the estimate of both the parameters of the Rasch model could be covered by the formula:

$$P(X_{ni} = 1|\beta_n, \delta_i) = \frac{e^{(\beta_n - \delta_i)}}{1 + e^{(\beta_n - \delta_i)}} \tag{10.5}$$

Where

- X_{ni} refers to response (X) elicited by the subject n to item i;
- β_n refers to the ability of subject n;
- δ_i refers to difficulty of item i;
- $X_{ni} = 1$ refers to a correct response of the item;
- e, indicates the base of the natural logarithm (i.e., $e = 2.718282$).

By applying the logistic model we were able to estimate the probability, for each subject, to partially or entirely solve the inferential problem and, conjointly, the probability associated to all the identified FCA solution's patterns. When the ability parameter β_n is higher than the related parameter δ_i the probability $P(X_{ni} = 1|\beta_n, \delta_i)$ of a correct response exceed 0.5; in the opposite case, the solution probability becomes smaller then 0.5. Both of the parameters estimated by the model identify a conjoint system of measurement of the collected dataset with respect to the psychological attitude of examinees and the complexity of the test to which they was exposed. The derived scale constitute a measure of a single latent trait.

It is now possible to introduce the specific application we carried out to test the applicability of the introduced elements.

2 Methods

2.1 Sample and Procedure

Participants were 256 students of the Psychometrics course 120 students were recruited at the University of Padua, while the remaining 136 students were recruited at the University of Cagliari. Students were asked to solve an "Hypothesis Testing for Paired Samples *t*-test" exercise. No time limit was imposed. Students were asked to solve the exercise in a paper and pencil way in order to allow the exact location of an error throughout the exercise. All participants attended the same theoretical and practice lessons on the topic during their Psychometrics course. Furthermore, both theoretical and practical lessons were carried out by the same teacher in both universities. Thus, given the fact that the sample size is almost the same for Padua and Cagliari, that the practical and theoretical lessons were exactly the same (same slides, same exercises, same text books) conducted by the same teacher, we can reasonably hypothesize the absence of substantial differences between the two groups.

The proposed exercise was initially subdivided into six main parts. After preliminary analysis through the Rash model, the following four parts were used to describe the whole exercise:

1. Hypothesis generation and formal expression;
2. Calculation of the test statistic t;
3. Identification of the critical value of test statistic t;
4. Assumption of the correct decision.

Each part of the exercise was evaluated independently from the others, i.e., a "wrong" conclusion coherent with the obtained calculated value of the t statistics and with the critical value of the statistic was right scored. The scoring procedure assume that a *local independence* for each part of the problem exists.

2.2 Analysis

The first part of the analysis involved the construction of the formal context which had the four parts of the exercise as the attributes and the response patterns as objects. Figure 10.1 displays the implications existing among different patterns with respect to the attributes dimension of the FCA solution.

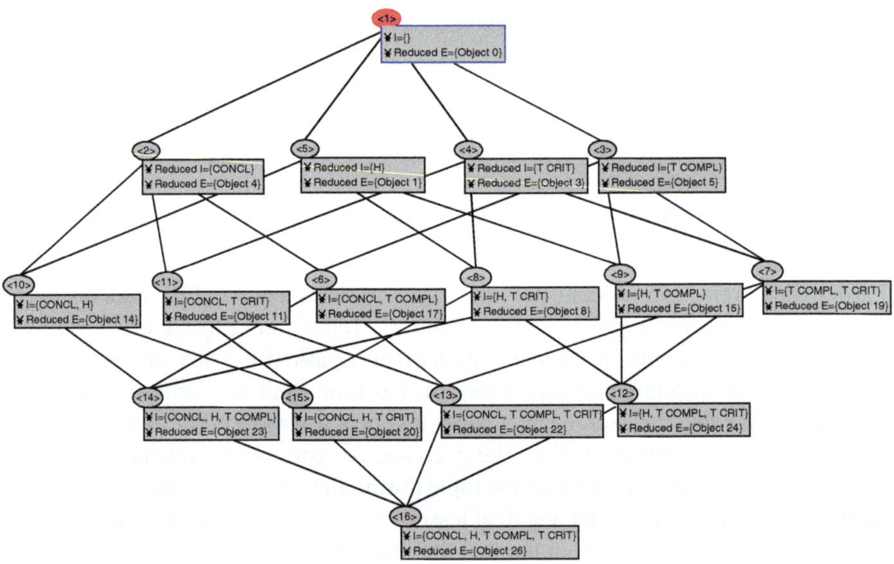

Fig. 10.1: The graphical representation of the implications among the parts of the exercise. In each node of the lattice a formal concept including a set of objects (response patterns) and a set of attributes (parts of the exercise) is represented. The four parts of the exercise are named as follows: H = hypothesis formulation and formal expression; T COMPL = calculation of the test statistic t; T CRIT = identification of the critical value of the test statistic t; CONCL = assumption of the correct conclusion

We then applied the Rasch simple logistic model to the context by means of both the RUMM 2020 and R statistical software. Through these analysis two main issues were obtained: on the one hand, it was possible to evaluate each part of the exercise in terms of its own difficulty (i.e. a location on the difficulty continuum was obtained for each item); on the other hand, through the evaluation of the residuals of the model, the ideal learning path moving from knowing nothing to knowing everything was found among different alternatives. The residual of a pattern indicates the discrepancy between the observed pattern and the expectations given by the model. The higher the residual, the greater is the discrepancy between the pattern and the model. Through this information it can be identified a path that minimizes the residuals, i.e. a set of steps that best represent the ideal learning process of a student.

3 Results

Table 10.1 displays both the location of the four parts of the exercise in terms of difficulty, and the residual analysis on the observed response patterns.

Table 10.1: The items' locations and the residual values of each response pattern

Items difficulty	Exercise part					Location
	I1: Hypothesis generation and formal expression					−0.422
	I2: Calculation of the test statistic t					0.402
	I3: Identification of the critical value of test statistic t					−0.223
	I4: Assumption of the correct decision					0.062
	Pattern/Item	I1	I2	I3	I4	Residual
	1	0	0	0	0	0.903
	2	0	0	0	1	−2.179
	3	0	0	1	0	−0.944
	4	0	1	0	0	−1.714
	5	1	0	0	0	−0.609
	6	0	0	1	1	0.594
	7	0	1	0	1	0.195
Residual analysis	8	1	0	0	1	−1.276
	9	0	1	1	0	1.254
	10	1	0	1	0	0.298
	11	1	1	0	0	−1.587
	12	0	1	1	1	0.774
	13	1	1	0	1	4.016
	14	1	0	1	1	0.086
	15	1	1	1	0	1.205
	16	1	1	1	1	−0.972

The higher the location, the more difficult is the item. The patterns are presented in a Boolean form in which a 1 in an entry means that the pattern includes the item

From the table it can be argued that the simplest part of the proposed exercise is the generation of the hypothesis and its formal expression, while the more difficult part is the calculation of the *t* statistics. It is noteworthy that the identification of the critical value of the test statistics appears to be less difficult than the assumption of the correct decision. With respect to the residual analysis, it can be highlighted that an ideal learning path includes the steps going from the empty set to the complete exercise through the mastering of item 1, then items 1 and 3, then items 1, 3 and 4. This path is included in the formal context and it is the one that both minimizes the residual of the model, and follows the difficulty order displayed by items locations.

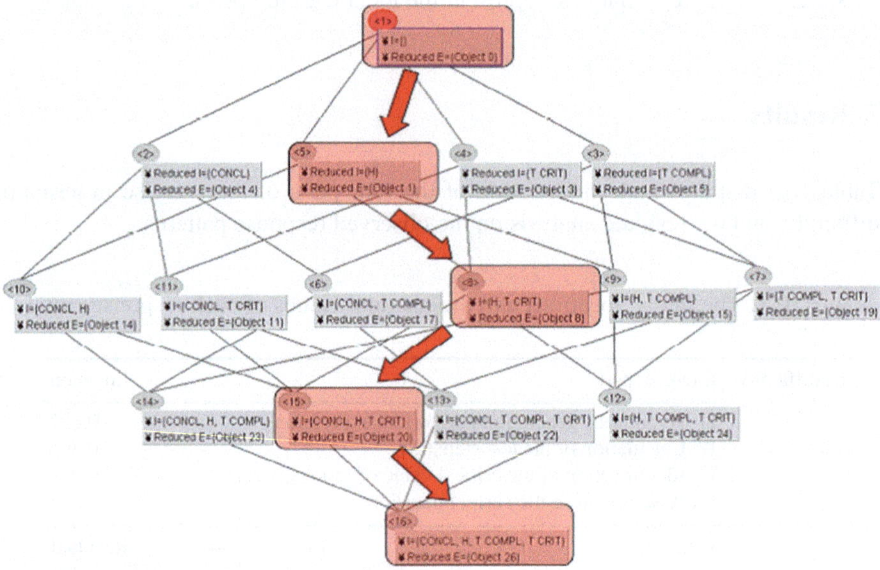

Fig. 10.2: The graphical representation of the results obtained through the analysis of the residuals. The highlighted nodes of the lattice represent the ideal learning path from the empty set (at the *top* of the figure), to the whole exercise (at the *bottom*)

Figure 10.2 displays the learning path depicted by the analysis of residuals. It can be seen how the path follows the increasing difficulty rate of the four parts of the exercise. By following the highlighted path the ideal sequence of concepts is learned. On the other hand, the present representation could be used as a reference point for calibrating an adaptive knowledge assessment tool. In fact, both the lattice and the path could be intended as the set representation of the implications existing among the parts of the exercise. Thus, if a student cannot formulate the correct hypothesis about the exercise, it would be almost certain that he/she will fail in computing the test statistic. The represented learning path is a total order. Through the FCA approach each partial order among items can be represented and implemented into an algorithm for the adaptive assessment of knowledge.

4 Discussion

The present work confirms the possibility of describing the solution process of a statistical exercise in terms of a number of steps that conduct from the total incapability to solve the exercise to a complete mastery of it. This process could be conveniently represented through the FCA and adequately described, from a probabilistic point of view, by the Rasch model. The use of FCA could play an important role in both algorithmically describe the solution process, and in planning specific teaching strategies accounting for the increasing difficulty of the proposed arguments and the logical sequence of exercise solution [5]. From the results it is possible to conclude that a reasonable solution to improve the ability of students to solve the proposed exercise may involve two main steps: the first one referring to the formal and logical part of the inferential process (hypothesis formulation, identification of the critical value of the test statistics, decision); while a second step of the learning process could be devoted to the more operational part of the exercise (i.e., the calculation of the value of the t statistics). The second part should follow the first one in order to maximize the probability of observing a correct answer to the exercise.

The presented approach could be fruitfully applied in adaptive knowledge assessment of students. More specifically, the implications obtained either a priori from theoretical consideration, or from a set of observed data, could be used to calibrate an algorithm that moves within the formal context by asking each time the more informative item. The procedure could allow to both improve assessment efficiency, and to precisely identify what exactly a student knows and what are his/her critical learning steps. Some software are already available for completing this task, but none of them refer to the FCA for detecting the specific relations among items [6–8].

Future works could investigate in more detail both the applicability of the method to other problem types [9] and to consider more sophisticated probabilistic models such as the partial credit [4].

References

1. Ganter, B., & Wille, R. (1999). *Formal concept analysis: Mathematical foundations*. Berlin: Springer.
2. Duquenne, V. (1999). Latticial structures in data analysis. *Theoretical Computer Science, 217*, 407–436.
3. Birkoff, G. (1967). *Lattice theory*. Providence, RI: American Mathematical Society Colloquium Publication.
4. Lord, F. M. (1980). *Applications of item response theory to practical testing problems*. Mahwah, NJ: Erlbaum.
5. Belohlavek, R., De Baets, B., & Konecny, J. (2014). Granularity of attributes in formal concept analysis. *Information Sciences, 260*, 149–170.
6. Schrepp, M. (1999). On the empirical construction of implications between bi-valued test items. *Mathematical Social Sciences, 38*, 361–375.

7. Schrepp, M. (2003). A method for the analysis of hierarchical dependencies between items of a questionnaire. *Methods of Psychological Research Online, 8*(1), 43–79.
8. Ünlü, A., & Malik, W. A. (2008). Psychometric data analysis: A size/fit trade-off evaluation procedure for knowledge structures. *Journal of Data Science, 6*, 491–514.
9. Xu, W., Pang, J. Z., & Luo, S. (2014). A novel cognitive system model and approach to transformation of information granules. *International Journal of Approximate Reasoning, 55*, 853–866.

Chapter 11
Emergence in Neural Network Models of Cognitive Processing

Maria Pietronilla Penna, Paul Kenneth Hitchcott, Maria Chiara Fastame, and Eliano Pessa

1 Introduction

The goal of this contribution is to assess whether the basic hypotheses underlying the connectionist approach are firmly grounded and useful for the research activities concerning the Psychology. These hypotheses can be summarized as follows:

1. the observed macroscopic consequences of cognitive processing (and, more in general, of mental processing) are nothing but collective effects emergent from the interactions between microscopic units; the latter operate on an observational level far lower than the one usually taken into consideration by experimental psychology;
2. hypothesis 1. putatively connects physico-chemical activities with mental processes; this in turn opens the way to a full solution of the mind-body problem;
3. the concrete implementation of these hypotheses is based on suitable mathematical models making use of *artificial neural networks* (hereinafter shortly called *neural networks*, or NN).

In order to reach our goal we must ascertain whether: (a) the NN models so far introduced concretely show emergent collective effects; (b) these collective effects are characterized by the same features which we observe in behaviors produced by human mental processes.

M.P. Penna (✉) · M.C. Fastame
Dipartimento di Pedagogia, Psicologia, Filosofia - Università di Cagliari, Cagliari, Italy
e-mail: penna@unica.it

P.K. Hitchcott
Department of Psychology, Southampton Solent University, Southampton, Hampshire SO14 0YN, UK

E. Pessa
Dipartimento di Scienze del Sistema Nervoso e del Comportamento - Università di Pavia, 27100 Pavia, Italy

G. Minati et al. (eds.), *Towards a Post-Bertalanffy Systemics*, Contemporary Systems Thinking, DOI 10.1007/978-3-319-24391-7_11

The arguments introduced in this paper, as well as the outcomes of numerical simulations performed in a case study led us to conclude that only particular models of NN can give rise to emergent collective effects. Moreover, despite the opportunities offered by these models, only the use of specific strategies and techniques of data analysis allows the use of the models themselves in a way useful to experimental psychologists. The conceptual road followed to reach these conclusions will be described in detail within the following sections. Finally the application to a specific case study will be used to illustrate the nature of the difficulties encountered when traveling along this road.

2 The Role of Experimental Data

In most cases the connectionist approach has been implemented by starting from general hypotheses about the phenomena or processes under consideration. Then, these hypotheses have been used to design a suitable NN model, with the hope that the behaviors of the latter be at least qualitatively similar to the ones already observed in the human beings. However, this strategy often is often ineffectual, a circumstance that has generated a widespread suspicion towards the connectionist models. Consequently most people do not understand why a sort of black box like a NN, consisting only in nodes and links, can help to account for complex human behaviors.

In this regard we claim that, in order to build a NN model with a reasonable hope of success, we should start from available experimental data. Only in this way can we, if our design strategy is effective, at least reproduce specific (and not generic) data. In the following, to simplify our arguments, we will take into consideration only data including *answers* of human subjects to suitable features of the experimental environment, collectively called *inputs*.

Once available a set of inputs associated with a specific sample of subjects, we can represent it as a set of *vectors* v_i $(i = 1, \cdots, N)$, in such a way that each vector describes the inputs associated with a particular subject. The components of each vector are given by the numerical values associated with the input features used in the experiment, whose number N is fixed and given in advance. We assume that, through suitable procedures, even non-numerical features can be represented in a numerical way. Thus in the following we will take into consideration only numerical vectors.

How to identify the different levels of description (macroscopic and microscopic)? According to our view, the macroscopic level is one of the outputs of the subjects (or of their statistical features). Instead, we will identify the single input features as the microscopic components of mental processes. However, as in the particular experiment taken into consideration the input features act in concomitance, we will choose input feature vectors, rather than single features, as micro-cognitive units (to use the connectionist language).

In this regard, we note that in most experiments the distribution of numerical values of input feature vectors is characterized by a considerable variance. This induced us to start the design of a NN model based on the available data, to begin with a *clustering* of the set of input vectors. Without entering into details on the clustering algorithm, we limit ourselves to point out that NN-based clustering methods (like, for instance, Kohonen's one) are highly convenient, both because they don't require artificial preliminary choices and, chiefly, because they identify immediately the vectors which represent the *prototypes* of the obtained clusters.

3 The Design of a NN Model Derived from the Data

The clustering procedure previously mentioned allows the association of each cluster with a *Radial Basis Function* neural unit (shortly a RBF unit). Each unit of this kind is a sort of detector sensitive to vector inputs very close to the prototype of the cluster associated with the unit itself. There are, as it is well known, many different kinds of RBF units. A typical activation law can have the form:

$$h_i = \frac{\exp[-(Y_i/\sigma_i)^2]}{\sum_{k=1}^{N_C} \exp[-(Y_k/\sigma_k)^2]} \tag{11.1}$$

where h_i is the output of the i-th RBF unit, N_C is the total number of clusters, σ_i is the width of the i-th cluster and Y_i is given by:

$$Y_i = \sqrt{\sum_{j=1}^{N} (x_j - P_j^{(i)})^2} \tag{11.2}$$

Here N is the feature number, x_j is the j-th component of the external input vector and $P_j^{(i)}$ is the j-th component of the prototype vector of the i-th RBF unit.

Having introduced the RBF units, the next step in the design of the NN model of experimental data would seem the construction of a multilayer network (a sort of Perceptron) in which the hidden layer includes the RBF units, the output layer codes the output features, and the input layer the input features. It would be natural, at this point, to introduce a supervised learning procedure based on the experimentally observed relationships between input and output features in such a way as to obtain a network able to associate to the input features received by a specific subject with it's experimentally observed outputs.

Unfortunately this strategy, even if feasible and probably successful, is not convenient if we want to reach goals (a) and (b) mentioned in the first section. Namely, supervised learning processes in the application consist of optimization algorithms designed to achieve goals well known and designed in advance. This issue prevents us from identifying these processes with emergence processes. Even if we introduce noisy aspects in these procedures, the study of the dynamical evolution of these networks, performed with the methods of Statistical Mechanics (see, for instance, [1, 5, 9, 10]), showed that this evolution coincides with a tendency towards stable equilibrium situations which, when the dimensions of the network and the data samples tend to infinity, are just the ones stated in advance by the designer.

If, on the contrary, we are searching for emergent effects we must abandon the domain of multilayer perceptrons and supervised learning, by focusing our attention on the so-called *recurrent* networks. These lack a subdivision into different layers and can therefore support, at least in principle, the occurrence of complex information feedback, allowing non-trivial dynamical evolutions and the occurrence of emergent collective effects. Moreover such networks have been used to build models of human behaviors, mostly in the domain of Social Psychology (see, for instance, [11, 12]). In recent times it has been shown (see [2, 6]) that recurrent networks with both excitatory and inhibitory synapses can undergo a transition from a state of global coherence to a state of uneven spatio-temporal distribution of activations. The latter is somewhat similar to the one of chaotic systems, which are highly sensitive to small differences in the external input. On the other hand, the eventual presence of chaos in recurrent networks is not a problem, owing to the fact that periodic inputs of sufficient amplitude (probable in networks receiving inputs from the environment or form other networks) can neutralize the effects of chaos (see [7]).

4 The Building of the NN Model

In agreement with the considerations made in the previous sections, the NN model would include a population of RBF unit containing as many members as the number of clusters obtained from experimental input data, as well as specific units deputed to receive the external input feature values and to send to the environment the output feature values. All these units should be connected through suitable recurrent connections whose weight values could be initially set at random. Some simple considerations show, however, that it is scarcely probable that such a kind of network can give rise to emergence effects. In particular:

1. the number of RBF units is expected to be very low, owing to the fact that clustering procedures cause a drastic reduction of information on the set of experimental input feature vectors; and we cannot classify effects produced by (only) a small number of units as emergent ones;
2. the experimental sample could not be representative of the whole universe of possibilities related to the phenomenon under study; thus, limiting ourselves to only those clusters obtained from the experimental data could give rise to a bias on the operation of a NN model which, instead, should be able to forecast behaviors, even if never encountered in the experimental sample used.

These considerations suggest the opportunity of widening the primeval recurrent network by artificially generating a bigger number of new prototypes and of the corresponding RBF units. In principle, this generation could be performed in a fully random way. If, however, we intend to fulfill criteria of psychological plausibility: the new prototypes should also be psychologically admissible variants of the ones experimentally found, even if never observed in actual experimental data. The new RBF units should then be connected through recurrent connections with the other units already present in the primeval model.

After the end of the building phase of this widened recurrent network, we can start a learning phase in order to find the most suitable values of the connection weights. Usually with recurrent networks learning is based on the so-called *delta rule*:

$$\Delta w_{ij} = \varepsilon(ext_i - \sum_k w_{ik}a_k)a_j \tag{11.3}$$

where Δw_{ij} denotes the weight change of the connection received by the i-th unit from the j-th one, ε is a parameter, ext_i is the external input to the i-th unit, the sum within parentheses concerns all units sending signals to this unit and a_j is the output activation of the j-th unit. This rule must be applied by sending to the input units the values of feature units occurring in the experiment and to the output units the corresponding values of output features experimentally observed.

The final NN model obtained after this learning procedure can then be studied in order to monitor its behavior in the presence of specific inputs and, chiefly, to study its dynamics in the search for the occurrence of emergent effects. In particular, if we could observe a dynamical evolution towards an attractor, the final phase of this evolution being characterized by the establishment of a dynamical coherence between the behaviors of a specific group of RBF units, then we could assert that the macroscopic cognitive aspect described by the attractor "emerges" from a collective coherence between the behaviors of a number of microscopic RBF units. In this way the main hypothesis of the connectionist approach would be confirmed by the NN model behavior and would acquire a well defined (and concrete) meaning. In order to assess the validity of our strategy we will present in the next section a case study showing the difficulties of the road to be covered to achieve our initial goals.

5 A Case Study

The case under consideration is given by an experiment on human performance in a serial order memory task. 144 subjects, each belonging to one of three different age groups (young, old, and very old) were required to recall earlier presented letter lists. In some cases the required recall modality was forward (the subject should recall the list items in the same order in which they appeared in the early presented list), in other cases backward (the should recall the items in reverse order with respect to the presented list). Some lists shared the same initial letter ("Start same" condition) while others shared the same final letter ("End same" condition). The performance measure of each subject was the total number of recall errors made in correspondence to all presentations. So far a report about this experiment is still unpublished. Thus, to obtain further details about the experiment, one must contact one of the authors of this paper. In any case, the short description given before is enough to understand and assess our modeling strategy.

Before starting our discussion, we remark that the domain of serial order memory is characterized by a number of still unexplained experimental effects, as well as by the lack of general quantitative models. As a consequence, it is very difficult to decide in advance what independent variables (among the many ones possible) should be monitored in order to account for all possible influences on subjects' performance within the chosen experimental context. Thus, the majority of experiments in this domain takes into account only a rather small number of independent variables. Our case study is not an exception, taking into consideration only five independent variables: the gender, the age, the standard score on the Mini Mental State Examination (MMSE), the recall modality, and the age group. Moreover, we had only three dependent variables: the total number of errors on recall of lists belonging to start same condition, the total number of errors on recall of lists belonging to end same condition, and the total number of errors in a control condition.

Initially, after a suitable recoding of variable values, in order to obtain numerical values suited to be used in a NN model, we chose the individual vectors of input values of 72 subjects and used this set for a clustering procedure. The latter was performed by a traditional Kohonen's network, with 5 input units and 20 categorization units (for details see a standard textbook on neural networks, such as, e.g., Rojas [8]). The used short-cut algorithm had a time dependence of the learning rate given by $\alpha = \eta e^{-\beta t}$ and of the bubble radius given by $r = r_0 e^{-b_0 t}$. The adopted numerical values for these parameters were: $\eta = 0.1$, $\beta = 0.0001$, $r_0 = 2$, $b_0 = 0.0001$. After 10,000 presentations of the 72 input vectors, we obtained that the network grouped them into 11 different clusters.

At this point we had at disposal 5 input units, 3 output units, and 11 RBF units, each one associated with the corresponding prototype vector obtained from the previous clustering, and with a variance given by the radius of the corresponding cluster. In order to build a primeval NN model we chose to connect in a random way all these units, by assigning to each link a weight randomly chosen from a Gaussian distribution with zero mean and variance given by $1/3$ for the connections pointing towards the output units, and $1/5$ for all other connections. The smallness of the number of RBF units induced us to insert 10 new RBF units, with randomly chosen prototypes and unit variances. However, to preserve psychological plausibility, each new unit was accepted with a probability given by $p = F(d - s)$ where d is the minimal Euclidean distance between its prototype and all other 11 initial prototypes, s is a suitable threshold parameter, and the function F is defined by $F(z) = \frac{1}{1-e^{-z}}$. Each one of the new RBF units was randomly connected to all previously existing units by a similar random procedure.

This enlarged NN was subjected to a learning procedure given by a suitable generalization of the delta rule already described in the previous section. More precisely the weight change was given by:

$$\Delta' w_{ij} = e^{-k\Delta t} \Delta w_{ij} \tag{11.4}$$

where Δw_{ij} is the weight change given by the traditional delta rule, Δt is the time interval between two successive learning steps, and k is a suitable parameter. This modification was introduced to prevent from the very fast growth of weight values produced by delta rule in very long learning procedures. In many usual conditions this is not a problem, as often the recurrent networks are used on very short time scales (see, e.g., Van Overwalle and Siebler [12]).

In our case the parameter values have been $\varepsilon = 0.1$, $D = 0.9$, $k = 3$. Here D is the spontaneous unit activation decay rate, entering into the law of unit activation update, given by:

$$a_i(t+\Delta t) = a_i(t) * External\ Input + \sum_j w_{ij} a_j(t) - D a_i(t) \tag{11.5}$$

This law did not apply to RBF units, whose activation law has been already described in the third section. It is to be remarked that, in correspondence to any pattern coming to the unit inputs of a RBF, the pattern itself was before normalized, in such a way as to have that the sum of the squares of input units activations was equal to 1. The learning was activated by the presentation to the network input units of all 144 vectors, one by one. Each input vector stayed for two steps. Moreover, the whole set of input vectors was presented twice, so that the total number of learning steps was $2 \times 2 \times 144 = 576$.

The next stage of our strategy consisted in studying a free evolution of the obtained NN model towards an attractor state, in absence of any learning, in order to monitor the occurring of emergence processes. However, the implementation of this stage required some changes in the obtained model. First of all, we have been forced to recognize that the random procedure of weight assignment could, in some cases, give rise to unwanted results. In our case, for instance, this procedure produced a consistent number of units endowed with output links but lacking input links. These units were clearly useless, being unable to influence the network evolution, and we have been forced to eliminate them. As this elimination caused a shortening of the available units, we tried to patch up the situation by introducing 30 new RBF units with randomly chosen prototypes. They were connected to all other units by links with random weights, having a probability of being excitatory of 0.5, and an absolute value (constrained to be within the interval from 1 to -1) chosen from a Gaussian distribution with mean 0.4 and variance 0.3. The new NN model obtained after these rearrangements had a total of 41 RBF units.

Other changes, however, have been required in order to have the possibility of a network time evolution characterized, on one side, by an attractor landscape enough complex and, on the other side, by a decay towards the attractor characterized by evolution times long enough as to allow a statistical analysis of the evolutionary features endowed with some meaning. In order to meet the former requirement we modified the updating rule of input and output units in the following way:

$$a_i(t+\Delta t) = F\left[a_i(t) + External\ Input + \sum_j w_{ij} a_j(t) - D a_i(t)\right] \tag{11.6}$$

where F denotes the function described above. As regards the RBF units we modified their activation law by adding to their output a random value chosen from a gaussian distribution with zero mean and variance 0.2. Then, by using a random initial pattern lasting for 5 initial time steps, we observed the free evolution of this new NN model for 100 steps.

We monitored the time evolutions of the single outputs of RBF units and, in order to analyse them through the methods commonly used in computational neuroscience, we recoded each output value as 1 if it was greater than 0.5 and as 0 in the contrary case. Thus, we obtained that the evolution of each RBF unit could be viewed as equivalent to a spike train. This allowed us to compute the cross-correlations between the activities of the different units and the *global coherence index* κ. The computation of the latter is based on a previous partitioning of the total evolution time into a succession of *time bins*, each equal to another. The amplitude of the single bin is chosen in such a way as to allow a meaningful statistical comparison between the concomitant activities of different units. Namely this comparison would be very difficult in the case on single time steps, owing to the fact that in most cases the instantaneous activations (within our recoding convention) could be zero. For our analysis we chose a bin amplitude of 5. Then, following the proposal of Wang and Buzsaki [13], we can define the coherence between two spike trains through:

$$\kappa_{ij} = \frac{\sum_{\ell=1}^{K} X(\ell)Y(\ell)}{\sqrt{\sum_{\ell=1}^{K} X(\ell) \sum_{\ell=1}^{K} Y(\ell)}} \tag{11.7}$$

where i and j are the indices of the two units, associated with the spike trains $X(\ell)$ and $Y(\ell)$, respectively. Instead, K denotes the total number of bin within the total evolution time. The global coherence index on a set of units, denoted by κ, is given by the average value of κ_{ij} on all possible unit pairs belonging to the set under consideration.

In the case of the total time evolution of our 41 RBF units the global coherence index was 0.5490609. We remind that the maximum possible value of this index is 1. However, when taking into consideration different stages of the network evolution, we observed different values of the coherence index. More precisely, when we deal with the evolution between the 20th and the 60th time step the coherence index was 0.5132665, while, when considering the evolution between the 60th and 100th step, the coherence index was 0.5600362. This difference, beyond signaling approach towards an attractor, suggested that we check whether the emergence of a collective cooperation between the activities of a subset of units could be responsible for this coherence increase. In this regard we were forced to deal with the problem of detecting the existence, within our network, of communities of units acting in a cooperative way (reminding old Hebb cell assemblies). Unfortunately this problem is very difficult to solve (see, e.g., Lancichinetti et al. [3]) and the absence, in our case, of a significant autocorrelation length and of meaningful Fourier coefficients does not help in this enterprise.

In order to find an approximate solution we resorted to the consideration that the members of a community of strongly correlated units should fulfill a principle according to which, if we have two units, denoted here as A and B, whose behaviors are strongly correlated, and two other units, denoted as B and C, equally strongly correlated, we should have that also A and C should be strongly correlated. Unfortunately the existence of the first two correlations does not always grant for the occurrence of the third. But a theorem (see Langford et al. [4]) asserts that, if the sum of the two first correlation coefficients is greater than 1, then surely also the third correlation is positive. In the case where the two first correlation coefficients are equal, the condition is satisfied when both correlation coefficients are greater or equal than $1/\sqrt{2}$. Then, by looking at the matrix of cross-correlations between the spike trains of the 41 RBF units and discarding all cross-correlations whose coefficients were lesser than $1/\sqrt{2}$, we were able to individuate 17 units meeting this condition. Computing now the global coherence index of this subset of units, we found it was 0.8821337, far larger than the one computed on the whole network. We are then justified in holding that we found proof for the existence, within our network, of a cell assembly made of units cooperating to obtain the output values characterizing the attractor of the evolution triggered by the used input. We could say that this output is a collective macroscopic effect emerging from the coherent activities of this subgroup of 17 RBF units.

It is interesting to observe that the prototypes associated with these units can suggest some theoretical considerations related to the experiment from which our strategy started. Namely, as the prototype vector components are normalized, their squares could be interpreted as giving the probabilities of influence associated with the psychological variables they are coding. In this regard, it is interesting to look at the mean values, on the whole set of 17 units, of the squares of the single prototype components. The relationships among these latter, even if their numerical value is meaningless, could give some indication about the relative importance of the independent variables. In our case we found that the most important variables are given by the gender (partially unexpected?) and the recall modality, together with a lesser importance of MMSE score. The age and the age group, on the contrary, do not seem to be important (strange enough, as the experiment was designed to compare the performance of older people with the one of younger people). These considerations, of course, suggest the need for a deeper analysis of the experimental results and, chiefly, of new experiments. In this way, an appropriate use of NN models shows that the latter could be useful to analyze the meaning of experimental data and to suggest new experimental investigations.

6 Conclusions

The previous arguments and the case study showed that, perhaps, the connectionist hypotheses have some grounding. However, in order to exploit the richness of the possibilities offered by NN models, we must choose, among them, only those permitting the occurrence of emergence processes. So far, the best candidates are offered by the recurrent networks, provided we adopt suitable evolution and learning laws. If we limit ourselves to the models of psychological process amenable to an experimental investigation, our strategy, besides being feasible, seems to offer a viable alternative for implementing the connectionist approach.

References

1. Coolen, A. C. C., Kühn R., & Sollich, P. (2005). *Theory of neural information processing*. Oxford: Oxford University Press.
2. Goudar, V., & Buonomano, D. V. (2014). Useful dynamic regimes emerge in recurrent networks. *Nature Neuroscience, 17*, 487–489.
3. Lancichinetti, A., Fortunato, S., & Kertész, J. (2009). Detecting the overlapping and hierarchical community structure in complex networks. *New Journal of Physics, 11*, 033015.
4. Langford, E., Schwertman, N., & Owens, M. (2001), Is the property of being positively correlated transitive? *The American Statistician, 55*, 322–325.
5. Mace, C. W. H., & Coolen, A. C. C. (1998). Statistical mechanical analysis of the dynamics of learning in Perceptrons. *Statistics and Computing, 8*, 55–88.
6. Ostojic, S. (2014). Two types of asynchronous activity in networks of excitatory and inhibitory spiking neurons. *Nature Neuroscience, 17*, 594–600.
7. Rajan, K., Abbott, L. F., & Sompolinsky, H. (2010). Stimulus-dependent suppression of chaos in recurrent neural networks. *Physical Review E, 82*, 011903, 1–5.
8. Rojas, R. (1996). *Neural networks. A systematic introduction*. Berlin/Heidelberg/New York: Springer.
9. Saad, D. (Ed.). (1998). *On-line learning in neural networks*. Cambridge: Cambridge University Press.
10. Sollich, P., & Barber, D. (1997). Online learning from finite training sets. *Europhysics Letters, 38*, 477–482.
11. Van Overwalle, F., & Labiouse, C. (2004). A recurrent connectionist model of person impression formation. *Personality and Social Psychology Review, 8*, 28–61.
12. Van Overwalle, F., & Siebler, F. (2005). A connectionist model of attitude formation and change. *Personality and Social Psychology Review, 9*, 231–274.
13. Wang, X.-J., & Buzsaki, G. (1996). Gamma oscillation by synaptic inhibition in a hippocampal interneuronal network model. *The Journal of Neuroscience, 16*, 6402–6413.

Chapter 12
Beyond Networks: Search for Relevant Subsets in Complex Systems

Andrea Roli, Marco Villani, Alessandro Filisetti, and Roberto Serra

1 Introduction

Most natural and artificial systems show some form of organization that can be modelled as a network of nodes. A prominent example of the success of this approach is the field known as complex networks science, which has achieved outstanding results in the analysis, design and control of complex network systems. Research in complex networks has so far mainly focused on the topological properties of the network that affect or explain the behavior observed. Nevertheless, in most interesting cases, the topology of the network might not provide sufficient information on system dynamics. For example, consider network models in which nodes are subject to nonlinear update functions: in this case, the actual relations among variables might not be captured by the topology, but they are rather well represented in terms of coordinated dynamical behavior of groups of nodes. Notable examples are Boolean networks [12], coupled map lattices [3] and functional connectivity graphs in neuroscience [9, 10]. Moreover, in several cases, the interactions

A. Roli (✉)
Department of Engineering and Computer Science (DISI), Alma Mater Studiorum University of Bologna, Via Venezia 52, I-47521 Cesena, Italy

European Centre for Living Technology, Ca Minich, S. Marco 2940, 30124 Venezia, Italy
e-mail: andrea.roli@unibo.it

M. Villani · R. Serra
Department of Physics, Informatics and Mathematics, University of Modena e Reggio Emilia, Via Campi 213b, 41125 Modena, Italy

European Centre for Living Technology, Ca Minich, S. Marco 2940, 30124 Venezia, Italy
e-mail: marco.villani@unimore.it; rserra@unimore.it

A. Filisetti
Current address: Explora s.r.l, Rome, Italy

European Centre for Living Technology, Ca Minich, S. Marco 2940, 30124 Venezia, Italy
e-mail: alessandro.filisetti@gmail.com

G. Minati et al. (eds.), *Towards a Post-Bertalanffy Systemics*, Contemporary Systems Thinking, DOI 10.1007/978-3-319-24391-7_12

among the interesting variables are largely, or at least partly, unknown; therefore, it is necessary to infer some hypotheses about the organization by observing system's behavior "from the outside". For these reasons, methods are required to go beyond network models, taking into account multiple and dynamical relations among nodes.

In this paper we address the issue of identifying sets of variables that are good candidates as "relevant subsets" (RSs) for describing the organization of a dynamical system. We will assume that some variables can be observed and that they change in time (possibly reaching an attractor state) and we will look for groups of variables that can represent the required RSs for describing the system organization. Note that the very notion of a RS is somewhat ill-defined. However this feature is shared by several other interesting concepts, like e.g. that of a cluster. In general, we stress that a good candidate RS (CRS for short) should provide important indications about some key features of the system organization. The main features of good CRSs can be tentatively identified as follows:

1. The variables that belong to a RS have a strong interaction and integration with the other variables of the same RS
2. They have a weaker integration with the other system variables or RSs

The outcome of the analysis we propose here is essentially a list of subsets, ranked according to the above criteria, that provide clues to understand the system organization. The list cannot be used *tout court*: its application requires some care, but it can lead one to discover very interesting non obvious relationships, above and beyond those provided by topological or structural information. Note also that RSs can overlap, i.e., they don't need to be disjoint sets.

A peculiar feature of the method that will be illustrated in the following section is that it does not require knowledge of the system structure and rules, as it only requires knowledge of the behavior in time of the variables $x_1 \dots x_N$, without any prior knowledge of the topology or of the dynamical rules. All that is needed is a set of numerical values of the relevant variables in time, therefore the method can be applied to models as well as to real-world data. Of course, if further pieces of information are available (e.g., concerning the topology), they can be profitably used.

The reason why our method does not require prior knowledge on system topology and rules is that it infers information on structure and dynamical relations among variables by exploiting techniques from information theory. While the idea of applying concepts and measures of information theory to the study of complex systems is certainly not new, we found that a measure of this kind, the Cluster Index—introduced by Tononi and Edelman [11] in the study of neural networks close to a stationary state—can be profitably generalized to study dynamical systems. This generalization is called here the Dynamical Cluster Index (DCI) and it has been introduced in [13]. This method has been successfully applied in uncovering the structure of artificial ad hoc systems (e.g., simple leader-followers models), of models of autocatalytic reaction sets, of genetic networks (e.g., the Arabidopsis Thaliana), of protein networks (e.g., the Mapk signalling system[1]) and robot controllers [5].

[1] Results presented in a contribution currently under review.

We present and discuss the DCI-based method in Sect. 2, while in Sect. 3 we illustrate the application of the CRSs search procedure to typical examples of Boolean networks. We conclude the contribution with Sect. 4.

2 The Dynamical Cluster Index

In this section we succinctly introduce the DCI. The interested reader can find more details in [2, 13].

Let us consider a system modelled with a set U of N variables assuming finite and discrete values. The cluster index of a subset S of variables in U, $S \subset U$, as defined by Tononi [11], estimates the ratio between the amount of information integration among the variables in S and the amount of integration between S and U. These quantities are based on the Shannon entropy of both the single elements and sets of elements in U.

According to information theory, the entropy of an element x_i is defined as:

$$H(x_i) = -\sum_{v \in V_i} p(v)\ log\, p(v) \tag{12.1}$$

where V_i is the set of the possible values of x_i and $p(v)$ the probability of occurrence of symbol v. The entropy of a pair of elements x_i and x_j is defined by means of their joint probabilities:

$$H(x_i, x_j) = -\sum_{v \in V_i} \sum_{w \in V_j} p(v,w)\ log\, p(v,w) \tag{12.2}$$

Equation (12.2) can be extended to sets of k elements considering the probability of occurrence of vectors of k values. In this work, we deal with observational data, therefore probabilities are estimated by means of relative frequencies.

The cluster index $C(S)$ of a set S of k elements is defined as the ratio between the integration $I(S)$ of S and the mutual information between S and the rest of the system $U - S$.

The integration of S is defined as:

$$I(S) = \sum_{x \in S} H(x) - H(S) \tag{12.3}$$

$I(S)$ represents the deviation from statistical independence of the k elements in S. The mutual information $M(S; U - S)$ is defined as:

$$M(S; U - S) \equiv H(S) - H(S|U - S) = H(S) + H(U - S) - H(S, U - S) \tag{12.4}$$

where $H(A|B)$ is the conditional entropy and $H(A,B)$ the joint entropy. Finally, the cluster index $C(S)$ is defined as:

$$C(S) = \frac{I(S)}{M(S;U-S)} \tag{12.5}$$

Since C is defined as a ratio, it is undefined in all those cases where $M(S;U-S)$ vanishes. In this case, the subset S is statistically independent from the rest of the system and it has to be analyzed separately. As $C(S)$ scales with the size of S, cluster index values of systems of different size need to be normalized. To this aim, a reference system is defined, i.e., the homogeneous system U_h, randomly generated according to the probability of each single state measured in the original system U. Then, for each subsystem size of U_h the average integration "I_h" and the average mutual information "M_h" are computed. Finally, the cluster index value of S is normalized by means of the appropriate normalization constant:

$$C'(S) = \frac{I(S)}{\langle I_h \rangle} / \frac{M(S;U-S)}{\langle M_h \rangle} \tag{12.6}$$

Furthermore, to assess the significance of the differences observed in the cluster index values, a statistical index T_c is computed:

$$T_c(S) = \frac{C'(S) - \langle C'_h \rangle}{\sigma(C'_h)} \tag{12.7}$$

where $\langle C'_h \rangle$ and $\sigma(C'_h)$ are the average and the standard deviation of the population of normalized cluster indices with the same size of S from the homogeneous system.

We emphasize that definitions (12.5)–(12.7) are made without any reference to a particular type of system. In their original papers, Edelman and Tononi considered fluctuations around a stationary state of a neural system. In our approach, this measure is applied to time series of data generated by a dynamical model. In general, these data lack the stationary properties of fluctuations around a fixed point. Moreover, depending upon the case at hand, either transients from arbitrary initial states to a final attractor, or collections of attractor states can be considered, as well as responses to perturbations of attractor states. In all these cases we will use Eq. (12.5), that will therefore be called the Dynamical cluster index (DCI).

The search for CRSs of a dynamical system by means of the DCI requires first the collection of observations of the values of the variables at different instants. It is not necessary to know the values of all the important variables, although of course the quality of the results can be negatively affected by lack of information. Moreover, in principle it is not required to have a time series: in fact, since the DCI is computed on the basis of symbol frequencies, a collection of snapshots of the system variables is sufficient.

In order to find CRSs, in principle all the possible subsets of system variables should be considered and their DCI computed. In practice, this procedure is feasible only for small-size subsystems in a reasonable amount of time. Therefore, heuristics are required to address the study of large-size systems. A simple heuristic consists in evaluating samples of subsets of increasing cardinality k: at first, subsets of size k are uniformly sampled and evaluated, then the samples of size $k+1$ are composed of random samples plus all the $(k+1)$-size neighbours of the k-size subset with the highest DCI value. This heuristic has proven to be quite effective, because usually the subsets with highest DCI value are composed of subsets which in turn have a high DCI value, compared to the subsets of the same cardinality. In our experiments, we always relied on the procedure described above. However, other search procedures can be adopted.[2] Once all the samples for size up to $N-1$ are evaluated, the T_c is computed so as to rank the CRSs.

3 Relevant Subsets in Boolean Networks

In this section we illustrate the application of the CRSs search procedure on networks whose nodes are updated according to Boolean functions, namely Boolean networks (BNs). BNs are an important framework frequently used to model genetic regulatory networks [4], also applied to relevant biological data [1, 7, 8] and processes [6, 12]. In the BNs we consider in the following, the nodes update dynamics is synchronous and deterministic, therefore every system state has only one successor state and the system trajectories in the state space can be decomposed into a transient and a cyclic attractor (which can also be a fixed point, i.e., a cyclic attractor of period 1).

The first example consists of a BN, named BN1, made of two independent components (see Fig. 12.1a).[3] The data used in order to compute the relative frequencies are obtained from the states of the various attractors, each one weighted according to its basin of attraction. The rationale for using the attractor states for this analysis is, intuitively, that attractors should be able to capture the important functional relationships in a system. The DCI analysis is able to correctly identify the two separated subnetworks of BN1 (data not shown for lack of space). This case simply provides an example of the concordance of the results provided by the DCI-based method and a topological clustering technique.

Example involving BN2 (Fig. 12.1b) is of great interest: as it can be observed, BN2 is composed of two interacting subnets: one composed of nodes {4,5,6} (identical to a BN module in BN1) and nodes {1,2}, which constitute an oscillator. The two subnets feed node 3, which make a XOR of nodes 1 and 6. If we analyzed BN2 in terms of its topology, we would split it into the three above-mentioned components. This analysis, though, does not capture the dynamical essence of the network, which instead involves all but one node (either node 1 or node 2 is excluded, as they are totally correlated), because they are all equally important to reconstruct the dynamical behavior of BN2. The DCI-based analysis returns one main CRS composed of nodes {2,3,4,5,6}, so it correctly clusters the relevant nodes for the

[2] As an ongoing work, we are experimenting with a genetic algorithm for searching the CRS with highest DCI value for each size.

[3] The following three examples has already been described in [13], however in this contribution we emphasize different aspects w.r.t. previous work.

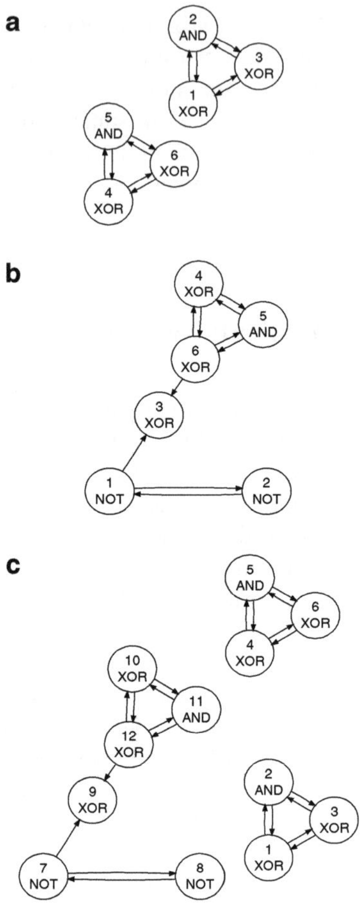

Fig. 12.1: (**a**) Independent Boolean networks (BN1); (**b**) interdependent networks (BN2); (**c**) a system composed of the union of both previous networks (BN3). The boolean function associated to each node is shown inside the node itself

dynamic of the system. Finally, the union of BN1 and BN2, named BN3, is depicted in Fig. 12.1c. An analysis based on the DCI correctly returns in the highest positions of the rank the two subnets of BN1 and the CRS of BN2.

One might argue that these examples are trivial, because they consider very small and simple BNs: on the contrary, while these BNs are simple if topology and rules are known, it is far from trivial to identify the most CRSs by the sole observation of the stationary states of the BNs.

In the following we present further notable cases of networks in which the topological structure is not sufficiently informative to detect the actual RSs. They

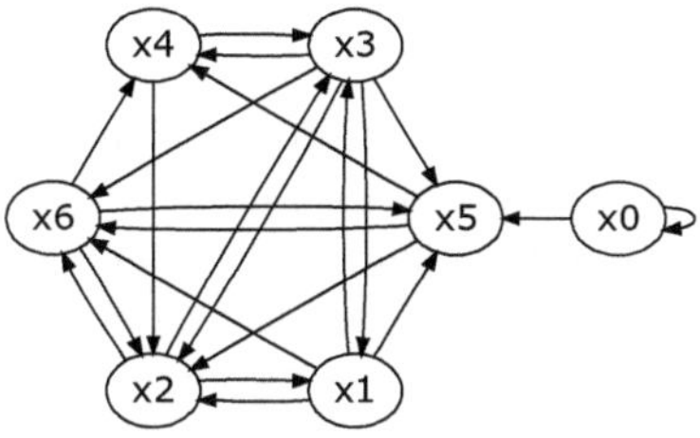


```
Node functions:

x0 = !x0
x1 = (x2 & !x3) | (!x2 & x3)
x2 = (x1 & !x3) | (!x5 & x3) |
     (x3 & x4) | (!x5 & !x6)
x3 = (x1 & !x2) | (!x4 & x2)
x4 = (x5 & !x3) | (!x6 & x3)
x5 = (x6 & !x1) | (!x1 & x3) & x0
x6 = (x1 & x3) | (!x2 & x5)
```

Fig. 12.2: BN with a topology composed of a cluster and a peripheral node, which indeed strongly influences the rest of the network

represent the not so rare case in which topology even defeats dynamical community identification. The first BN under exam is depicted in Fig. 12.2, along with node update functions. The network has 5 attractors: two fixed points, 2 cycles of period 2 and one of period 14. A topological analysis would identify two clusters, namely $\{x_0\}$ and $\{x_1,\ldots,x_6\}$. Nevertheless, a dynamical analysis made by means of the DCI returns as first CRS the subset $\{x_0,x_1,x_2,x_4,x_5,x_6\}$, showing that from the dynamical point of view node $x_0$is not disconnected from the rest of the network. This is a prominent example of the situation in which an agent external to a (topological) community can anyway influence it.

An example related to defeating community detection techniques based on pairwise relations is provided by the BN depicted in Fig. 12.3. The BN has 3 attractors of period 4, 5 and 13, respectively. Despite the typical structure displaying two loosely connected clusters, the dynamics of the network is such that nodes x_1 and x_6 are indeed strongly interdependent and the DCI-based analysis shows that the most CRSs involve nodes from both the topological clusters. For example, among the first five suggested CRSs, the largest one is composed of 8 nodes, all except nodes x_2 and x_4.

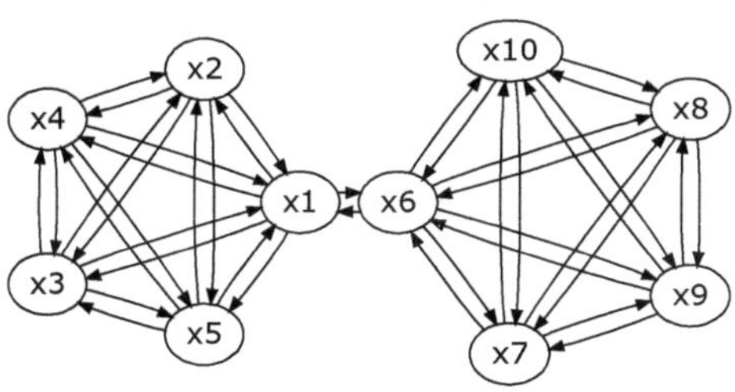

Fig. 12.3: BN whose topology is composed of two cliques (node update functions are omitted for lack of space). The actual dynamics, though, is such that the nodes of the two clusters are strongly integrated

4 Conclusion

In this contribution we have presented a method, based on information theory, which makes it possible to detect CRSs of variables in a system, exploiting the data deriving from the dynamics of the system. Although the advantage of the method is to identify CRSs emerging from the dynamics of the system and it does not require prior knowledge on system's structure, a promising direction for future work consists in combining the topological information with the dynamical one.

Acknowledgements The research leading to these results has received funding from the European Union's Seventh Framework Programme (FP7/2007-2013) under grant agreement n^{0} 284625.

References

1. Balleza, E., Alvarez-Buylla, E., Chaos, A., Kauffman, S., Shmulevich, I., & Aldana, M. (2008). Critical dynamics in genetic regulatory networks: Examples from four kingdoms. *PloS One, 3*(6), e2456.
2. Filisetti, A., Villani, M., Roli, A., Fiorucci, M., Poli, I., & Serra, R. (2014). On some properties of information theoretical measures for the study of complex systems. In: C. Pizzuti and G. Spezzano (Eds.), *Advances in artificial life and evolutionary computation*, CCIS 445, pp. 140–150, Springer.
3. Kaneko, K. (1990). Clustering, coding, switching, hierarchical ordering, and control in a network of chaotic elements. *Physica D, 41*, 137–172.
4. Kauffman, S. (1993). *The origins of order: Self-organization and selection in evolution.* Oxford: Oxford University Press.
5. Roli, A., Villani, M., Serra, R., Garattoni, L., Pinciroli, C., & Birattari, M. (2013). Identification of dynamical structures in artificial brains: An analysis of boolean network controlled robots. In *AI*IA2013: Advances in Artificial Intelligence. Lecture notes in computer science* (Vol. 8249, pp. 324–335). New York: Springer.
6. Serra, R., Villani, M., Barbieri, A., Kauffman, S., & Colacci, A. (2010). On the dynamics of random Boolean networks subject to noise: Attractors, ergodic sets and cell types. *Journal of Theoretical Biology, 265*(2), 185–193.
7. Serra, R., Villani, M., Graudenzi, A., & Kauffman, S. (2007). Why a simple model of genetic regulatory networks describes the distribution of avalanches in gene expression data. *Journal of Theoretical Biology, 246*, 449–460.
8. Serra, R., Villani, M., & Semeria, A. (2004). Genetic network models and statistical properties of gene expression data in knock-out experiments. *Journal of Theoretical Biology, 227*, 149–157.
9. Shalizi, C., Camperi, M., & Klinkner, K. (2006). Discovering functional communities in dynamical networks. In *Proceedings of ICML 2006. Lecture notes in computer science* (Vol. 4503, pp. 140–157). New York: Springer.
10. Sporns, O., Tononi, G., & Edelman, G. (2000). Theoretical neuroanatomy: Relating anatomical and functional connectivity in graphs and cortical connection matrices. *Cerebral Cortex, 10*(2), 127–141.
11. Tononi, G., McIntosh, A., Russel, D., & Edelman, G. (1998). Functional clustering: Identifying strongly interactive brain regions in neuroimaging data. *Neuroimage, 7*, 133–149.
12. Villani, M., Barbieri, A., & Serra, R. (2011). A dynamical model of genetic networks for cell differentiation. *PloS One, 6*(3), e17703.
13. Villani, M., Filisetti, A., Benedettini, S., Roli, A., Lane, D., & Serra, R. (2013). The detection of intermediate-level emergent structures and patterns. In *Advances in Artificial Life, ECAL 2013* (pp. 372–378). New York: The MIT Press.

Chapter 13
From Elementary Pragmatic Model (EPM) to Evolutive Elementary Pragmatic Model (E^2PM)

Piero De Giacomo, Rodolfo A. Fiorini, and Giulia F. Santacroce

1 Introduction

The Elementary Pragmatic Model (EPM) was developed in the 1970s [9, 31, 37], following Gregory Bateson's constructivist participant observer concept in the "second order cybernetics", to arrive to what was called "new cybernetics", according to cybernetics classical historical categorization. EPM is a high operative and didactic, versatile tool and new application areas are envisaged continuously [14]. Quite recently, EPM intrinsic Self-Reflexive Functional Logical Closure contributed to find an original solution to the dreadful double-bind problem in classic information and algorithmic theory (i.e. our contemporary systemic tools and classic information computational and communication algorithms are totally unable to discriminate the difference between an optimal encoded information-rich message and a random jumble of signs that we call "noise" usually) [13, 15]. In turn, this new awareness has allowed to enlarge our panorama for neurocognitive system behavior understanding, and to develop information conservation and regeneration systems in a numeric self-reflexive/reflective evolutive reference framework [13]. Accordingly, new methods and models to build effective applications and strategies, from new forms of inter- and trans-disciplinarity, can be conceived conveniently. Rational human thinking is like a solid archipelago emerging out of an ocean of unaware intuitions. Human brain is an harmonization machine fed by unaware intuitions to produce learning and rational awareness about our environment. Our "Eulogic Thought" emerges out of

P. De Giacomo
Department of Neurological and Psychiatric Sciences, University of Bari, Bari, Italy

R.A. Fiorini (✉)
Department of Electronics, Information and Bioengineering (DEIB), Politecnico di Milano University, 32, Piazza Leonardo da Vinci, 20133 Milano, Italy
e-mail: rodolfo.fiorini@polimi.it

G.F. Santacroce
BV-Tech (S.p.A), Milano, Italy

G. Minati et al. (eds.), *Towards a Post-Bertalanffy Systemics*, Contemporary Systems Thinking, DOI 10.1007/978-3-319-24391-7_13

a continuous harmonization interaction between Paleologic and Neologic Thought components. "Emotional Intelligence" (EI) and "Emotional Creativity" (EC) [19] coexist at the same time with "Rational Thinking", (RT) sharing the same input environment information. Different forms of knowledge representations inducing a non-deductive procedure during inferences exist. For instance, the metaphor is a representation modality that bypasses deductive procedure. The source of the metaphor brings a structure exhibiting explicit and implicit attributes of the target. The metaphor allows direct inferences and learning about new things (target) by extending what it is already known about the source [1, 2, 24]. Similarly, the process of hierarchical inheritance in semantics networks allows direct inferences on properties shared by classes, subclasses and instances of objects without effort of explicit deduction [6]. Another form of direct inference is heuristics that take advantage of specialized knowledge to permit inferential short-cuts [4, 18]. Mental imagery also provides with detailed representation of objects that conveys implicit information without deductive effort [11]. In logic, diagrammatic notation is the main form of knowledge representation that excludes a deductive procedure through the process of spatial inference. EPM provides us with a reliable closed logic starting scheme to face "unknown known" situations [16, 17]. If we like to use it on more and more complex applications with ability to capture natural emergent phenomena dynamics, we need to extend it to be able to face even unpredictable perturbation ("unknown unknown") at design level [16, 17]. In this case, an evolutive structure to manage unexpected dynamics is needed to EPM. To cope with ontological uncertainty effectively at system level, it is possible to use two coupled irreducible information management subsystems, based on the following ideal coupled irreducible asymptotic dichotomy: Information Reliable Predictability and Information Reliable Unpredictability. In this way, to behave realistically, overall system must guarantee both Logical Closure (Reactive Information Management, "to learn and prosper") and Logical Aperture (Proactive Information Management, "to survive and grow"), both fed by environmental "noise" (better … from what human beings call "noise") [13]. So, a natural operating point can emerge as a new Trans-disciplinary Reality Level, out of the Interaction of Two Complementary Irreducible Information Management Subsystems. Building on this idea, it is possible to envisage an Evolutive Elementary Pragmatic Model (E^2PM) able to profit by both classic EPM intrinsic Self-Reflexive Functional Logical Closure and new evolutive Self-Reflective Functional Logical Aperture.

2 EPM Logical Structure

EPM was conceived as a Boolean model of two binary inputs x_0, P, and x_1, Q, for three distinct elements. In two-valued logic there are 2 nullary operators (constants), 4 unary operators, 16 binary operators, 256 ternary operators, and, in general, 2^{2^n} n-ary operators. Classically the domain and range of a truth function are {truth, falsehood}, but they may have any number of truth values, including an infinity of these. A concrete function may be also referred to as an operator. EPM "elementary interaction coordinates" (unary operators) are $2^2 = 4$, and they were named: (001) Antifunction, (011) Acceptance, (101) Maintenance and (111) Sharing. So, there are sixteen possible truth functions (binary operators), also called Boolean functions for two inputs. Any of these functions corresponds to a truth table of a certain logical connective in classical logic, including several degenerate cases such as a function not depending on one or both of its arguments. The arity of a function or operation is the number of arguments or operands the function or operation accepts. The arity of a relation (or predicate) is the dimension of the domain in the corresponding Cartesian product. A function of arity "a" thus has arity $n = a + 1$, if considered as a relation. Usually, truth and falsehood is denoted as 1 and 0. An example of Boolean operations is illustrated in full, for $a = 2$ in Fig. 13.1. Finally, EPM can accommodate $2^8 = 256$ ternary operators. Please, note, that in this case, those 256 values can be thought as EPM power set $P(W)$, where $W = 2^3 = 8$. $P(W)$ can be even interpreted as EPM intrinsic Self-Reflexive Functional Logical Closure.

Therefore, EPM associated Boolean algebra is B_3. The Boolean algebra B_3 can be represented as a cube C_3 in three-dimensional Euclidean space R^3. This is done by the "conventional" Left-To-Right (LTR) coordinate mapping $c : \{0;1\}^3 \to R^3$, (Fig. 13.2).

It is well-known that a Boolean algebra can always be visualized by means of a Hasse diagram that is centrally symmetric (with all complementary pairs of elements ordered around the center of symmetry) [8]. Furthermore, a finite Boolean algebra can be partitioned into "levels" L_0, L_1, L_2, which are recursively defined as follows: $L_0 = [\perp]$, and

$$L_{k+1} = \{x \mid \exists y \in L_k : y \lhd x\}. \tag{13.1}$$

x_0	x_1	$^2f_0$	$^2f_1$	$^2f_2$	$^2f_3$	$^2f_4$	$^2f_5$	$^2f_6$	$^2f_7$	$^2f_8$	$^2f_9$	$^2f_{10}$	$^2f_{11}$	$^2f_{12}$	$^2f_{13}$	$^2f_{14}$	$^2f_{15}$
0	0	0	1	0	1	0	1	0	1	0	1	0	1	0	1	0	1
1	0	0	0	1	1	0	0	1	1	0	0	1	1	0	0	1	1
0	1	0	0	0	0	1	1	1	1	0	0	0	0	1	1	1	1
1	1	0	0	0	0	0	0	0	0	1	1	1	1	1	1	1	1

Fig. 13.1: Example of Boolean operations of arity 2

Hasse diagrams for the Boolean algebra $\phi(\{1,2,3\})$, and for a Boolean algebra of formulas from the modal logic $S5$ are shown in Fig. 13.3. It is immediate to map cube C_3 to Hasse diagram on left side of Fig. 13.3. Boolean algebras are locally finite and their word problem is always decidable (closed logic).

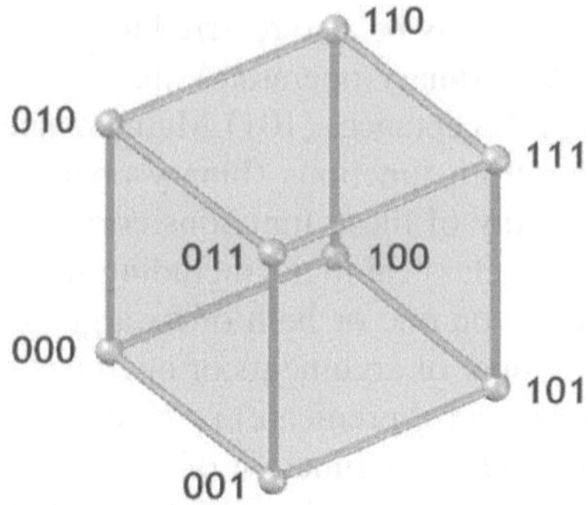

Fig. 13.2: EPM associated Boolean algebra B_3 is represented (LTR) by cube C_3 with its bitstring decoration in R^3

3 Extending EPM as a Boolean-Valued Model

To provide EPM with an evolutive structure (open logic), we can follow many different approaches. In previous section, we saw that EPM associated Boolean algebra B_3 can be represented LTR by cube C_3 in R^3. By remembering the notions of "logical space" proposed by Wittgenstein [43] and of "hypercube" proposed by Pólya [34], it is straightforward to consider a Boolean-valued model, as the simplest extension of EPM. A Boolean-valued model is a generalization of the ordinary Tarskian notion of structure from model theory [41]. In a Boolean-valued model, the truth values of propositions are not limited to "true" and "false", but instead take values in some fixed complete Boolean algebra. Boolean-valued models were introduced by Dana Scott, Robert M. Solovay, and Petr Vopěnka in the 1960s in order to help understand Paul Cohen's method of forcing, presented in 1963 [5]. They are also related to Heyting algebra semantics in intuitionistic logic [42]. The problem of whether a given equation holds in every Heyting algebra was shown to

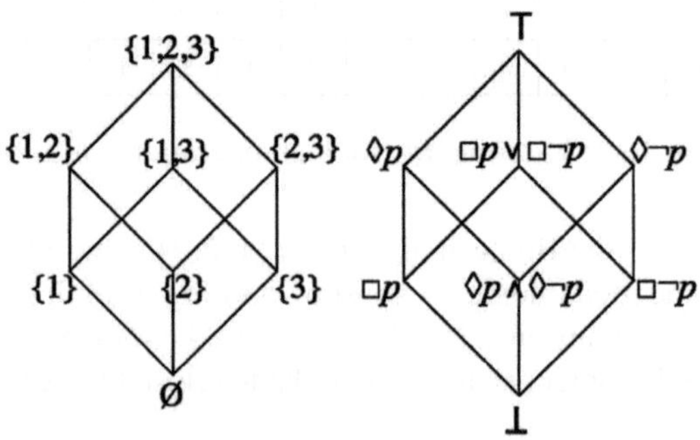

Fig. 13.3: Hasse diagrams for the Boolean algebra $\phi(\{1,2,3\})$ on the left side and for a Boolean algebra of formulas from the modal logic $S5$ with modality $(\Box, \Diamond)$, on the right side [10]

be decidable by S. Kripke in 1965 [28]. The precise computational complexity of the problem was established by R. Statman in 1979, who showed it was PSPACE-complete [38] and hence at least as hard as deciding equations of Boolean algebra (shown NP-complete in 1971 by S. Cook [7]) and conjectured to be considerably harder. The elementary or first-order theory of Heyting algebras is undecidable [20]. It remains open whether the universal Horn theory of Heyting algebras, or word problem, is decidable [26]. For the word problem it is known that Heyting algebras are not locally finite (no Heyting algebra generated by a finite nonempty set is finite), in contrast to Boolean algebras which are locally finite and whose word problem is decidable. It is unknown whether free complete Heyting algebras exist except in the case of a single generator where the free Heyting algebra on one generator is trivially completable by adjoining a new top. A Boolean algebra of order 2^n, called B_n, is graded of rank n, [29] and can be represented as a hypercube or n-cube C_n, in n-dimensional Euclidean space R^n, for $n = 0, 1, 2, 3, 4, \cdots, \infty, n \in N$. Our main idea is to achieve EPM open logic model behavior (logic dynamics) by providing EPM with the asymptotic process of the structured sequence of locally finite Boolean algebras for $n \to \infty$ theoretically. In Fig. 13.4, the process to obtain successive n-dimensional hypercubes up to $n = 4$ is depicted. For $n = 4$, the Boolean algebra is represented as a four-dimensional hypercube C_4. We again can employ the "conventional" LTR coordinate mapping $c : \{0;1\}^4 \to R^4$. This can be generalized to any number of dimensions. In fact, the process of sweeping out volumes (Fig. 13.4) can be formalized mathematically as a Minkowski sum: the n-dimensional hypercube is the Minkowski sum of n mutually perpendicular unit-length line segments, and is therefore an example of zonotope. Then, n-dimensional hypercubes geometrical information can be projected to convenient projection planes to study the local behavior of their connection components as graphs. For instance, in Fig. 13.5, the related Petrie polygon Orthographic projections up to $n = 8$ are shown. Then, the graph of the n-hypercube's edges is isomorphic to the Hasse diagram of the $(n-1)$-simplex's face lattice. This can be seen by orienting the n-hypercube so that two opposite vertices lie vertically, corresponding to the $(n-1)$-simplex itself and the null polytope, respectively.

Each vertex connected to the top vertex then uniquely maps to one of the $(n-1)$-simplex's facets ($n-2$ faces), and each vertex connected to those vertices maps to one of the simplex's $n-3$ faces, and so forth, and the vertices connected to the bottom vertex map to the simplex's vertices. This relation may be used to generate the face lattice of an $(n-1)$-simplex efficiently, since face lattice enumeration algorithms applicable to general polytopes are more computationally expensive.

Then, the Hasse diagram for any B_n can be seen as $(n-1)$-dimensional vertex-first projections of these hypercubes. Although Hasse diagrams are simple as well as intuitive tools for dealing with finite posets, it turns out to be rather difficult to draw "good" diagrams. The reason is that there will in general be many possible ways to draw a Hasse diagram for a given poset.

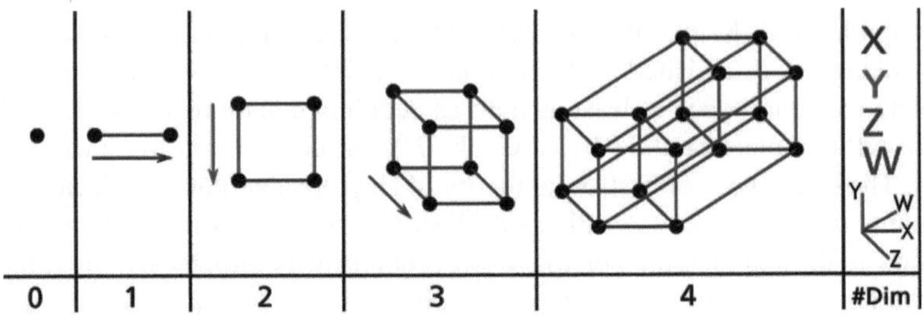

Fig. 13.4: (*0*)—A point is a hypercube of dimension zero. (*1*)—If one moves this point one unit length, it will sweep out a line segment, which is a unit hypercube of dimension one. (*2*)—If one moves this line segment its length in a perpendicular direction from itself; it sweeps out a two-dimensional square. (*3*)—If one moves the square one unit length in the direction perpendicular to the plane it lies on, it will generate a three-dimensional cube. (*4*)—If one moves the cube one unit length into the fourth dimension, it generates a four-dimensional unit hypercube (a unit tesseract)

The simple technique of just starting with the minimal elements of an order and then drawing greater elements incrementally often produces quite poor results: symmetries and internal structure of the order are easily lost. In Fig. 13.6 this issue is demonstrated by considering the power set of a 4-element set ordered by inclusion ($\subseteq$) of a 4-cube or tesseract. There are four different Hasse diagrams for this partial order. Each subset has a node labelled with a binary encoding that shows whether

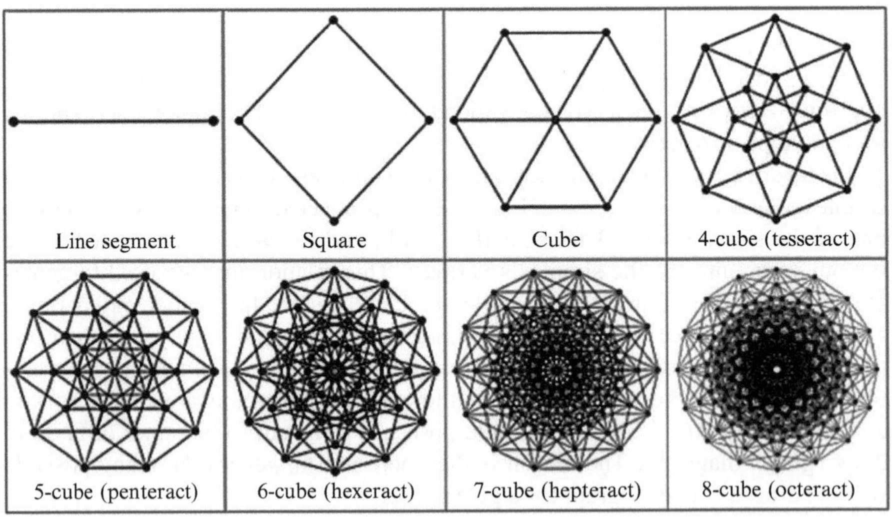

Fig. 13.5: N-dimensional Hypercube Petrie polygon Orthographic projections from $n = 1$ up to $n = 8$, as graphs

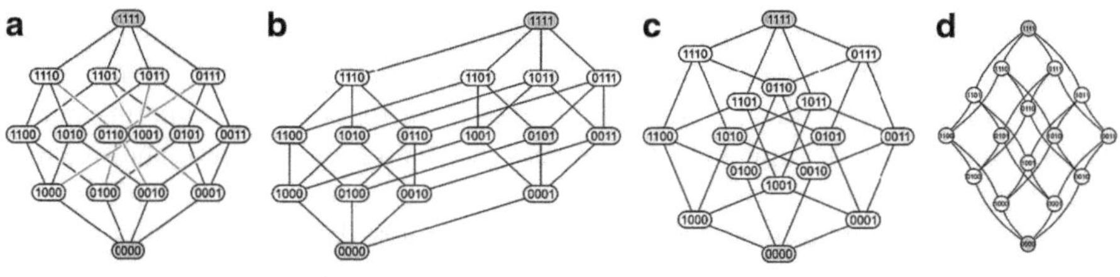

Fig. 13.6: Hasse diagrams (see text). The (**a**) first diagram on the left makes clear that the power set is a graded poset. The (**b**) second diagram has the same graded structure, but by making some edges longer than others, it emphasizes that the four-dimensional cube is a union of two three-dimensional cubes. The (**c**) third diagram shows some of the internal symmetry of the structure. In the (**d**) fourth diagram, on the right, the vertices are arranged like the fields of a 4×4 matrix

a certain element is in the subset (1) or not (0). Therefore, whether the diagram is Hasse or Aristotelian depends on our choice of the projection axis [10, 32]. For a full definition of these structures, by a theoretical perspective, one should look to Judson, at least [27]. Finally, please, note, that the power set $P(B_n)$ of any locally finite Boolean algebra B_n can be thought even as a Self-Reflective Functional Logical Closure for the power set $P(B_{n-1})$ of preceding locally finite Boolean algebra $P(B_{n-1})$. According to brand new Computational Information Conservation Theory (CICT), this property is fundamental to achieve overall model systemic resilience and antifragility behaviour) [13, 16, 17].

3.1 Ontological Uncertainty Modeling

To behave realistically (i.e. to capture natural event dynamics), system must guarantee both Logical Closure (to get RT, to learn and prosper) and Logical Aperture (to get EI and EC, to survive and grow), both fed by environmental noise (better … from what human beings call "noise") [13]. All the while almost everything, classically approached and traditionally studied, about social life focuses on the "normal", particularly with "bell curve" methods of inference that tell you close to nothing about natural events. Why? Epistemic uncertainty sources are still treated with the traditional approach of risk analysis only, which provides an acceptable cost/benefit ratio to producer/manufacturer, but in some cases it may not represent an optimal solution to end user [16, 17]. In fact, deep epistemic limitations reside in some parts of the areas covered in decision making. In fact, the bell curve ignores large deviations, cannot handle them, yet makes us confident that we have tamed uncertainty [40]. On the other end, almost everything in social life is produced by rare but consequential shocks and jumps. As the experiences of the 1970s, 1980s, 1990s and 2000s have shown, unpredictable changes can be very disorienting at enterprise level. These major changes, usually discontinuities referred to as fractures in the environment rather than trends, will largely determine the long-term future of organization. They need to be handled, as opportunities, as positively as possible. Model developers must concentrate on not ignoring or double counting uncertainties and clearly documenting the process in which they represent and quantify uncertainties. Traditionally, uncertainties are characterized as epistemic, if the modeler sees a possibility to reduce them by gathering more data or by refining models. Uncertainties are categorized as aleatory if the modeler does not foresee the possibility of reducing them. From a pragmatic standpoint, it is useful to categorize the uncertainties within a model, since it then becomes clear as to which uncertainties have the potential of being reduced. Influences of the two types of uncertainties in reliability assessment, codified design, performance-based engineering and risk-based decision-making are always present. More importantly, epistemic uncertainties may introduce dependence between events, which may not be properly noted if their character is not correctly modeled. Unfortunately, decision theory, based on a "fixed universe" or a model of possible outcomes (closed logic), ignores and minimizes the effect of events that are "outside model". In fact, human made systems can be quite fragile to unexpected perturbation because Statistics can fool you [39]. While the advantage of differentiating between natural (aleatoric) and epistemic uncertainty in analysis is clear, the necessity of distinguishing between them is not, at operational or operative level. As a matter of fact, epistemic and aleatory uncertainties are fixed neither in space nor in time. What is aleatory uncertainty in one model can be epistemic uncertainty in another model, at least in part. And what appears to be aleatory uncertainty at the present time may be cast, at least in part, into epistemic uncertainty at a later date [21]. It is much better to consider ontological uncertainty [30] as an emergent phenomenon out of a complex system [16, 17]. Then, our ontological perspective can be thought only as an emergent, natural operating point out of, at least, the interaction dichotomy of two coupled irreducible complementary ideal asymptotic concepts: (a) reliable predictability (closed logic subsystem) and (b) reliable unpredictability (open logic subsystem). From top-level management perspective, the reliable predictability concept can be referred to traditional system reactive approach (fixed logic subsystem) and operative management techniques, while the reliable unpredictability concept can be associated to system proactive approach (open logic subsystem) and strategic management techniques (Fig. 13.7).

4 Concluding Remarks

Now, traditional EPM can be thought as a reliable starting subsystem (closed logic, operative management, Fig. 13.7) to initialize a process of continuous self-organizing and self-logic learning refinement (open logic, strategic management subsystem, Fig. 13.7). As already described in previous sections, this method can capture natural logic dynamics behavior, as function of specific unpredictable perturbation, unknown at system design level. Though the hypercube logical geometry

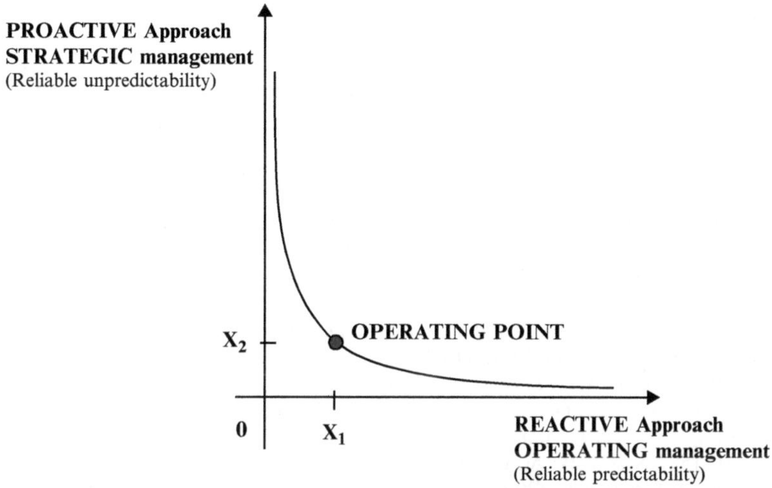

Fig. 13.7: Operating Point can emerge as a new Trans-disciplinary Reality Level, based on the Interaction of Two Complementary Irreducible Management Subsystems

seems to be a straight-forward method to depict the logical relations in propositional logic, further research must be planned to go beyond this first approach of the notation. Future studies ought to validate empirically the contribution of this logical geometry approach to the understanding of logical relationships, notably in educational settings. The intuitive character of the related algebra to apprehend logical relations must be tested in comparison with classical methods of learning. Through the hypercube geometric algebra, we propose a notation that goes beyond a format distinction and constructed with the purpose to facilitate inferences either on a diagrammatic representation, or a lexical one. The latter particularly allows operations on complex propositions within hypercube with more than three dimensions, mentally difficult to imagine. This algebra, by posting directly configurations in which a complex proposition is true, can explicitly represent all mental models, in the sense of Johnson-Laird [25], necessary for the apprehension of a proposition in all its complexity. In agreement to Morineau [33], we think that this algebra could represent a tool for assisting work activities that involve inductive reasoning, like problem- and case-based reasoning in medical diagnosis [12], and subject profiling in psychiatry and psychotherapy [3, 22, 23]. More specifically, from a biomedical engineering perspective, fault diagnosis task [35] and troubleshooting on logical networks [36] could be areas of application for reliable testing and validation of the presented EPM extension as "Evolutive Elementary Pragmatic Model" (E^2PM). E^2PM presents a relevant contribute to models and simulations offering an example of new forms of evolutive behavior inter- and trans-disciplinarity modeling.

References

1. Bateson, G. (1991). Men are grass: Metaphor and the world of mental process. In R. E. Donaldson (Ed.), *A Sacred Unity: Further steps to an ecology of mind* (pp. 235–242). New York: A Cornelia and Michael Bessie Book.
2. Blackwell, A. (2001). *Thinking with diagrams*. Dordrecht: Kluwer Academic Publishers.
3. Chang, G., Weiss, A. P., Orav, E. J., Smallwood, J. A., Gonzalez, S., & Kosowsky, J. M., et al. (2012). Bottlenecks in the emergency department: The psychiatric clinicians' perspective. *General Hospital Psychiatry, 34(4)*, 403–409.
4. Cheng, P., & Holyoak, K. (1985). Pragmatic reasoning schemas. *Cognitive Psychology, 17*, 391–416.
5. Cohen, P. J. (1966). *Set theory and the continuum hypothesis*. Boston: Addison-Wesley.
6. Collins, A. M., & Quillian, M. R. (1969). Retrieval time from semantic memory. *Journal of Verbal Behavior and Verbal Learning, 8*, 240–247.
7. Cook, S. A. (1971). The complexity of theorem proving procedures. In *Proceedings of 3rd Annual ACM Symposium on the Theory of Computing* (pp. 151–158). New York: ACM. doi:10.1145/800157.805047.
8. Davey, B., & Priestley, H. (2002). *Introduction to lattices and order*. Cambridge: Cambridge University Press.
9. De Giacomo, P., Silvestri, A., Corfiati, L., Lefons, E., Pierri, G., Pazienza, M. T., et al. (1981). Versuch einer interaktiven "logischen" Therapie. Ein theorisches Modell fuer Therapie und Bevertung der Veraenderung. *Familiendynamik, 2*, 148–157.
10. Demey, L., & Smessaert, H. (2014). The relationship between Aristotelian and Hasse diagrams. *Diagrammatic Representation and Inference, 8578*, 213–227.
11. Denis, M. (1979). *Les images mentales*. Paris: Presses Universitaires de France.
12. Eshach, H., & Bitterman, H. (2003). From case-based reasoning to problem-based learning. *Academic Medicine, 78*, 491–496.
13. Fiorini, R. A. (2014). How random is your tomographic noise? A Number Theoretic Transform (NTT) approach. *Fundamenta Informaticae, 135*, 135–170.
14. Fiorini, R. A., De Giacomo, P., Marconi, P. L., & L'Abate, L. (2014). Quality of care as an emergent phenomenon out of a small-world network of relational actors. In J. Mantas (Ed.), *Proceedings of 12th International Conference on Informatics, Management and Technology in Healthcare*, Athens, 10–13 July 2014.
15. Fiorini, R. A., & Laguteta, G. (2013). Discrete tomography data footprint reduction by information conservation. *Fundamenta Informaticae, 125*, 261–272.
16. Fiorini, R. A., & Santacroce, G. F. (2014). A post-Bertalanffy systemics healthcare competitive framework proposal. In J. Mantas (Ed.), *Proceedings of 12th ICIMTH International Conference on Informatics, Management, and Technology in Healthcare*, Athens, 10–13 July 2014.
17. Fiorini, R. A., & Santacroce, G. F. (2014). Safety and effectiveness health systemic governance by HICT natural framework. In F. Pinciroli (Ed.), *Proceedings of International Conference on Biomedical Informatics*. Milano, 8–9 May 2014.
18. Gigerenzer, G. (2000). *Adaptive thinking: Rationality in the real world*. Oxford: Oxford University Press.
19. Goleman, D. J. (1995). *Emotional intelligence: Why it can matter more than IQ*. New York: Bantam Books.
20. Grzegorczyk, A. (1951). Undecidability of some topological theories. *Fundamenta Mathematicae, 38*, 137–152.
21. Hanks, T. C., & Cornell, C. A. (1994). Probabilistic seismic hazard analysis: A beginner's guide. In *Proceedings of the Fifth Symposium on Current Issues Related to Nuclear Power Plant Structures, Equipment and Piping* (pp. I/1-1–I/1-17). Raleigh, NC: North Carolina State University.
22. Hillard, R., & Zitek, B. (2004). *Emergency psychiatry*. New York: McGraw-Hill.

23. Høglend, P. (2014). Exploration of the patient-therapist relationship in psychotherapy. *American Journal of Psychiatry*. doi: 10.1176/appi.ajp.2014.14010121 [Epub ahead of print]
24. Indurkhya, B. (2006). Emergent representations, interaction theory, and the cognitive force of metaphor. *New Ideas in Psychology, 24*, 133–162.
25. Johnson-Laird, P. N. (2001). Mental models and deduction. *Trends in Cognitive Science, 10*, 434–442.
26. Johnstone, P. T. (1982). *Stone spaces*. Cambridge: Cambridge University Press.
27. Judson, T. (2013). *Abstract algebra: Theory and applications*. GNU Free License Document (7th ed.). Accessed 30 July 2014. http://abstract.pugetsound.edu
28. Kripke, S. A. (1965). Semantical analysis of intuitionistic logic I. In J. N. Crossley & M. A. E. Dummett (Eds.), *Formal systems and recursive functions* (pp. 92–130). Amsterdam: North-Holland.
29. Kusraev, A. G., & Kutateladze, S. S. (1999). *Boolean valued analysis*. Berlin: Springer.
30. Lane, D. A., & Maxfield, R. R. (2005). Ontological uncertainty and innovation. *Journal of Evolutionary Economics, 15*, 3–50.
31. Lefons, E., Pazienza, M. T., Silvestri, A., Tangorra, F., Corfiati, L., & De Giacomo, P. (1977). An algebraic model for systems of psychically interacting subjects. In *Ination and Systems. Proceeding of I.F.A.C. Workshop*, Compiègne, France. Oxford: Pergamon Press.
32. Moretti, A. (2009). *The geometry of logical opposition*. Ph.D. thesis, University of Neuchatel, Switzerland.
33. Morineau, T. (2012). Hypercube algebra: A diagrammatic and sentential notation to support inferences in logic. In *Proceedings of the 30th European Conference on Cognitive Ergonomics*, ECCE '12 (pp. 100–104).
34. Pólya, G. (1940). Sur les types des propositions composées. *Journal of Symbolic Logic, 5*, 98–103.
35. Rouse, W. B. (1978). Human problem solving performance in a fault diagnosis task. *IEEE Transactions on Systems, Man, and Cybernetics, SMC, 8*, 258–271.
36. Sanderson, P. M., & Murtagh, J. M. (1989). Troubleshooting with an inaccurate mental model. In *Proceedings of IEEE International Conference on Systems, Man, and Cybernetics* (pp.1238–1243).
37. Silvestri, A., & De Giacomo, P. (1979). A computerized laboratory for approach to relational psychotherapy. In *Proceedings of Congress on Medical Informatics*, Berlin.
38. Statman, R. (1979). Intuitionistic propositional logic is polynomial-space complete. *Theoretical Computer Science, 9*, 67–72.
39. Taleb, N. N. (2005). *Fooled by randomness* (2nd ed.). New York: Random House & Penguin.
40. Taleb, N. N., & Goldstein, D. G. (2012). The problem is beyond psychology: The real world is more random than regression analyses. *International Journal of Forecasting, 28*(3), 715–716.
41. Tarski, A. (1986). In S. R. Givant & R. N. McKenzie (Eds.), *The collected papers of Alfred Tarski* (4 Vols.) Berlin: Birkauser.
42. Troelstra, A. S. (Ed.). (1973). *Metamathematical investigation of intuitionistic arithmetic and analysis*. Berlin: Springer.
43. Wittgenstein, L. (1922). *Tractatus logic-philosophicus*. New York: Keagan Paul.

Part V
New Forms of Inter- and Trans-disciplinarity

Chapter 14
Systemic Approach and Meaningful Complexity in Biology

Mirko Di Bernardo

1 Dynamic Instability and Natural Order

In 1963 an event took place that soon would help to significantly change the way we look at reality in all the sciences, including physics: Lorenz discovered deterministic chaos, whose foundations were laid in 1889 by Poincaré with the three-body problem. Lorenz showed how in order to have a chaotic behavior of a dynamical system, a very simple model of nonlinear differential equations was sufficient. In this case, in fact, despite the strict determinism of Newton's law, one is faced with chaotic system behavior caused by the extreme sensitivity of the solutions of the equations to the initial conditions. So, it happens that two states, as similar to each other as you like, will distance themselves (become dissimilar) from each other exponentially over time. From the impossibility, not just practical, but in principle, to define the initial conditions with infinite precision, there descends therefore a substantial system state unpredictability that becomes less and less able to be predicted with the growing interval of time elapsed from the initial instant. Behold, then, the concept of deterministic chaos, a type of chaos that involves the fact that an exponential increase in knowledge of the present is required to maintain a significant contact with the past and future evolution of the system [30]. We find the root of randomness and unpredictability not in external reality or in the subject treated as separate domains, but in the persistent relationship that exists between the time of evolution for specific activities of the source and of the cognitive agent. From an effective point of view, chaos essentially mixes state space trajectories. The process of "stretching-folding" leads to a generalized dispersion of the points over the whole attractor, rendering any kind of forecast impossible. Chaos, then, is randomness, but a randomness that has a deterministic basis, which is linked to the defining of a very precise setup, which

M. Di Bernardo (✉)
Department of Business Government Philosophy, University of Rome "Tor Vergata", Via Columbia 1, Rome 00199, Italy
e-mail: dbrmrk01@uniroma2.it

G. Minati et al. (eds.), *Towards a Post-Bertalanffy Systemics*, Contemporary Systems Thinking, DOI 10.1007/978-3-319-24391-7_14

has elements of regularity and that is characterized by a coupled game of external constraints and fluctuations [8]. Chaos, in other words, is a systematic construction of a collective of Bernoullian stamp within which are expressed at the same time, according to specific conditions, principles of invariance and pure irregularities. It is, in other words, a random process that came to articulate itself with respect to one or more specific, contextual constraints, that intervene on a random process already underway [9]. Thus in all disciplines of the 1960s and 1970s (parallel to the studies of Monod) new languages were born, suitable to represent the properties of systems characterized by a functional and structural complexity that prevents one from deducing what they are from those of their constituents. They are based on the insufficiency of reductionism as the only valid scientific method, accepting the irreducibility of the various different levels of organization of such systems and the impossibility to find comprehensive explanations of their properties without resorting to historical and evolutionary categories (biological organisms, mind, social organization, economies). According to the school of Brussels the phenomena of irreversibility and self-organization rest upon a well-defined microscopic base. The basic idea that underlies the work of this school is that irreversibility is closely linked to the notion of dynamic instability. In the prediction of the behavior of unstable systems, in fact, it is not our lack of knowledge that is at play, but the dynamic nature of the system [31]. Therefore, dynamic instability is at the origin of the concept of probability and not vice versa. To clarify the meaning of this statement simply remember how, for Prigogine [33], by subjecting a particular type of system to a given constraint, we can obtain as a result an increase in entropy that is related, at the same time, to the emergence of a phenomenon of order [34]. The mechanism underlying this type of phenomenon is, essentially, an amplification mechanism of the fluctuations. Far from equilibrium there is an amplification of fluctuations that opens the way for a series of varied possibilities. Non-equilibrium thermodynamics deals with systems that have exchanges with the environment, systems in which the change in entropy is related not only to processes that occur within the system, but also to the flows of matter and energy between the system and environment. In this type of system, the decisive quantity is no longer the entropy, but the production of entropy, the entropy change per unit time with respect to the processes taking place within the system [8]. As is well known, since 1967, Prigogine defines these systems "dissipative structures", or structures that are a form of supermolecular organization. In these systems, therefore, unlike equilibrium thermodynamic systems where the balance is associated with the fall towards the most likely and least ordered state, the flow of matter and energy constitutes a driving force that generates order.

Applying this theory to biology by following the thoughts of the Russian scholar, you can infer that the structures adapt to external conditions, showing a kind of preadaptation mechanism and that, in conditions of being far from equilibrium, the matter begins to be able to perceive differences in the outside world (such as gravitational or electrical fields); this might not make any sense whatsoever in terms of balance, because matter is blind to balance [35]. From this perspective life appears much less opposed to the normal laws of physics, much less in a fight against them to prevent its normal fate, which would be its destruction. On the contrary, life seems to express somehow precisely the conditions in which it is immersed our biosphere, if account is taken of all the nonlinearities of chemical reactions and conditions of distance from equilibrium that solar radiation imposes on the biosphere [36].

2 The Systemic Vision

During work on the laws of chaos Prigogine and others showed how Boolean networks are logical-mathematical models (though limited) of a large class of non-linear dynamical systems [32]. Attractors of these networks can simulate the natural object of interest. From a biological point of view, according to Kauffman [22], one can hypothesize that these attractors match the cell types, while from a cognitive point of view it is possible to interpret these attractors as the natural classification that the network does of the outside world. These findings represent a prudent widening of the results obtained in the field of non-equilibrium thermodynamics. In particular it is important to point out, in this regard, that this enlargement concerns, first of all, the nature and dynamics of differentiation processes, the link, in perspective, which exists between these processes and the subsequent formation of particular basins of attraction. In such circumstances, therefore, in my opinion, it is possible to affirm that Kauffman, using the languages of Dynamics to interpret the biological phenomena, develops a mathematical model (plausible from a biological point of view) to enter the mystery of ontogeny in a broader theoretical framework in which biology suddenly finds itself in "dialogue" with other skills such as mathematics, physics, chaos theory, computer science and systems theory. Well, in this context, through the theory of randomly built Boolean networks, the American biochemist responds definitively to the need of Theoretical Biology Club (represented by the line of research begun by Waddington [38] with studies on canalization and genetic assimilation) to build a new organicistic non-vitalist paradigm of development in which biology is endowed with that power of logical and mathematical explanation which the physical sciences have always had. With this in mind, then, Kauffman, giving a mathematical code and a new method of approach to biology, offers, at the same time, an effective model able to give body to the original theorizing that overcame the classical dichotomy between mechanism and vitalism and which Waddington, in the 1940s, defined as the third systemic way, which can be summarized schematically in the following points: (a) life is a phenomenon that is not solely determinable by the physical and chemical laws to which nevertheless it is bound; (b) neither is there a special property of life, an intangible ingredient that directs its course; (c) the secret of functioning of living systems is the layering of evolutionary levels, irreducible to one another yet interacting; (d) the passage from one level to another corresponds to the succession of emergent properties, produced by interactions between the different evolutionary units of each level; (e) the living object in its entirety is given by its morphological and functional organization; (f) this organization is in a vital relationship at the same time of continuity and autonomy in relation to the physical principles [39].

In light of all this, then, the systemic vision of Waddington, in my opinion, can be summed up, in agreement with the results obtained by Prigogine and Kauffman, as follows: life is an emerging phenomenon that develops when the molecular diversity of a prebiotic chemistry system exceeds a certain level of complexity. If this is true, then life is not located in the individual property of each individual molecule, but is a collective property of systems of molecules interacting with each other. From this perspective, life emerged as a whole and has always remained a whole. In the whole that emerges and self-reproduces there is no vital force or foreign substance.

> But the collective system does possess a stunning property not possessed by any of its parts. It is able to reproduce itself and to evolve. The collective system is alive. Its parts are just chemicals [23, p. 24].

In this way, therefore, in transition from the theory of dissipative structures to the theory of self-organized Boolean networks we can actually perceive the development of a coherent and continuous research aimed at identifying the general principles that characterize the deep reality of that mysterious self-organization that marks the complexity of the *bios*.

3 Meaningful Complexity, Self-organization and Biological Information

The understanding of deep processes of self-organization concerning biological systems requires today a systemic and interdisciplinary approach. As highlighted by recent studies in the framework of an extended theory of Meaningful Complexity [9, 17, 20], there is more to the simple examination of Markovian-style dissipative phenomena, but one comes to consider the phenomena of the coupled processing and transformation of information present at the level of the subsequent constitution of a biological system characterized by information processing itself. From this context, from the end of the 1980s, some studies have tried, first of all, to illuminate the inner biological articulation, at the biological and cognitive level, of that particular process constituted by the emergence of teleonomical and intentional structures that underlie the processes of life in an evolutionary and co-evolutionary framework [28, 40]. This emergence is essentially tied to precise procedures for transmission and assimilation of in-depth information that arise, however, at the semantic and probabilistic level and that determine the subsequent articulation of a specific biological "code" considered as an operative synthesis of function and meaning, that arises as an effective support for the dynamic constitution of precise teleological structures [4, 5, 10]. Here we can recognize with accuracy that particular interweaving of complexity, self-organization, intentionality and emergence of meaning that characterizes the natural forms of cognitive activity of any living system. The analysis on the results obtained by the Human Genome Project, in fact, both reductionism and naive holism (already refuted in 1953 with the discovery of the DNA Double Helix) are definitely outweighed by a new theoretical synthesis which does deal with the idea of emergence of meaning in which the parts interact with each other and the whole giving rise to systemic circularity wider than that outlined for example, by Waddington and the early Kauffman. What characterizes the bios, therefore, is no longer only teleonomy as conceived by Monod (design without intention), but there begins to emerge the conception according to which life is inextricably linked to the idea of meaning, intentionality and memory (from the cell, to the immune system, to the apparatuses, all the way to the mind). According to this new interpretative framework which addresses the emerging qualities, life not only appears to be tied to a program written in the double helix, but, above all, is linked to the circularity of distributed programs related to self-programming, or the idea of biological meaning [14, 15]. The genetic information of the organism does not reside in the initial conditions of the dynamic process of ontogeny, but in distributed programs that govern new information and make it impossible, given the initial conditions, to predict with certainty the final state of the organism in question. Today we are aware of the hidden mechanism that allows the DNA, through the genetic code, to control the synthesis of proteins: the dynamics of auto-programming, in fact, is the same functionality of the genome that creates the genetic information [27]. The secret of self-organization that escaped Monod, which we identified in the concept of biological meaning, can be, in other words, identified in that creative function which generates the syntax (the genetic information of DNA nucleotides) and which is the basis of life [3, 18]. The biological meaning, in my opinion, is the "hidden face" of genetic information, the creating and organizing function that responds to a mathematics which are, in many aspects, as of yet unknown, a mathematics, for example, of the infinite that goes beyond Cantor's theorem and Kolmogorov's complexity theory, which could explain those highly complex phenomena and currently not completely explained, unpredictable and non-measurable by human reason using only statistic rarity or computational incompressibility. We have attributed the term deep reality to this foundation of life that exists, but is not (at the moment) understood, a reality that escapes biology, mathematics, physics and chemistry and that, however, allows us to study the life as an emerging phenomenon as a free and unyielding order [13]. The concept of Monodian invariance is now being revisited in the light of the emergence of meaning which, surpassing genetic determinism, completes it. So the mathematical modeling (Markovian processes and Boolean algebra) that allowed the first [22] to interpret the evolution of dynamic systems and stochastic processes of gene expression are no longer sufficient since we no longer find ourselves before a simple Markovian stochastic automaton, but are faced with non-standard models of complex cellular automata able to channel the flow of energy in such a way as to enable the hidden potential in this same stream to be progressively revealed in always different ways. In fact, as Atlan [3, 4] notes correctly, in a natural system that self-organizes, the end is not established from the outside. What self-organizes is the function itself with its meaning. The origin of meaning in the organization of the system is, therefore, an emergent property. In addition, the origin of meaning is closely related with precise options of observation. If we plan to build a complex cellular network, in order to simulate the activities of a biological system (cognitive) must take account of the fact that the behavior of the network has any significance not only to the extent that the result will be autonomous, but also in so far as the result is observed and intentionally tied to continuous production of new possible interpretations [5]. So, for an information source to be able to show independent behavior that will self-organize, we must add to these processes of mutation, selection, and special differentiation also the ability of observation, self-observation, simulation and interpretation. From an objective point of view, it must be noted, first, that the boundary between order and chaos seems to be able to offer much more sophisticated tools to selection [2, 8]. In particular it is able to offer, rather than point mutations, a wide variability, able to lead the environment to deploy itself by manifesting its hidden potential at a deep level. With reference to this particular "landscape" the constraints imposed by the selective pressures at the level of the dynamics of dissipative cellular automaton can actually allow a more complex channeling of the inbound information flow [11]. Through the development of the process now outlined, the genome determines, therefore, the progressive construction of a specific channel for another partial expression of deep informational content and revelation of new forms of incompressibility. Thus, the source of the information, in order to achieve a stable form of expression (a new order) must encapsulate itself in specific generative properties. These properties must be included in the physical matter of the system so as to give rise to the possibility of generating the produced varied complexity of the properties themselves [12].

4 Towards a Semantics of Biological Processes

In this theoretical framework, therefore, it seems clearly that to realize the old TBC project related to the construction of a theoretical biology independent from chemistry and physics, the observations, remodulations and abstract design regarding a statistical mechanics of a renewed nature as identified and pursued by Kauffman in *The Origins of Order*, and repeatedly revisited in his later texts, do not suffice. To build at the biological level a statistical mechanics concerning genes and macromolecules (in action) it is necessary to reckon fully with profound information, an information, namely, not measurable through the use of tools offered by the traditional Shannon information theory based on a mathematics that is too simple and thus "incompatible" with the complexity of vital phenomena. We must, in other words, define, as we previously mentioned, the principles of a new algorithmic information theory (i.e. a new complexity theory), not exclusively linked to a propositional basis, but articulated on the level of a logical dimension of a predicative and stratified character. Such a theory of complexity should be able, among other things, to show us how it is possible to speak, without any contradiction, of non-existence of finite algorithms in relation to issues which are well placed in terms of uniqueness and existence (non-existence is a necessary matter of departure just as, on the physical side, in agreement with Prigogine, the existence of randomness that is rooted in dynamics is a primitive given). Well, this also implies the elaboration of an intensional and hyper-intensional semantics for recurrent processes of self-organization, and the construction of automata simulation models endowed with intensional bases and reflexive and interpretive functions. In this sense, the first stage of a project so vast should be to refer, at least from an abstract point of view, to the attempts underway to delineate new conceptual principles to define an appropriate logical background for a correct semantics of biological processes [7, 11, 37]. These efforts have focused so far on the provisional definition of at least two new central concepts: the truth considered not as invariance but as emergence, and the model that self-organizes. As regards the first we are no longer faced with a notion of truth as a simple form of invariant propagation within the frame of a monotonic logical structure. The truth seems to now be defined, as Carsetti notes correctly, only by reference to non-monotonic procedures, at the level of the second order, to the actions of coupled systems, to the existence of specific intensional functions. In this sense the emergence of truth seems to be specifically linked not only to the processes of revelation and articulation of the original information source, but above all to the preliminary distinction between surface information and depth information [12, 21]. What characterizes, however, even better, that particular revolution paradigm in the field of semantics of our time represented by the actual delineation of an adequate semantic of processes, is perhaps the general concept of the model that self-organizes inherently, associated with dynamic aspects of the concept of meaning. At this level we are no longer just linked to the existence of individuals and logical invariants forms, as in the case of the theoretical constructions of Russell, Tarski and Henkin; we are, on the contrary, linked, first, to the articulation of specific and complex attributes of generators, that is to say, to the existence of attractors, operational closures, recurring flows of information and, in general, to a multi-tiered architecture of self-organization [24, 25]. And it is precisely with regard to this particular type of logic that the domains of individuals-object can then begin to articulate their existence, seeming, on the logical level, to be the end result of the effective articulation of some specific processes, in particular an internal building process and an effective functional partition process, where information flows represent "the true fibre" of the structure of the dynamic model itself [12]. Therefore, we are once again faced with some of the brilliant insights of Prigogine. According to the Russian scholar, in fact, to explain irreversibility (and stochasticity) you must consider states with a temporal symmetry-breaking propagated through laws which are themselves due to a break of symmetry [26]. The temporal symmetry breaking, therefore, in that context, represents an essential tool for developing a new level of understanding in which rationality is no longer identified with the idea of certainty. Similarly, therefore, it is possible to say that in the semantics of processes we witness the gradual introduction of concepts related to particular conditions of symmetry breaks that occur logically and informationally, such as, for instance, the concepts of partition and of self-organizing models. However, if on the other hand also by virtue of an intuition of Prigogine today we can penetrate new territories of semantics, in a terrain, namely, that appears to be strictly determined by the progressive expansion of evolutionary processes [6, 29]; on the other hand it should be noted once again how the current level of semantics, as well as to the level of a theory of multidimensional information, we are no longer confined within the limits of simple

Markovian frames. This fact constitutes a real dividing line with respect to the formal apparatus developed by Nicolis and Prigogine at the level of their exploration of the mathematical basis of the theory of complexity [1]. Within the framework of process semantics, therefore, we really need to resort to the delineation of new and more complex informational spaces, of new measures (new axioms) of complexity that can express themselves not only propositional-level (as, for example, the Shannon entropy), but also at the level of the first and second order [8]. It is only with reference to these more complex informational spaces that these functional partition processes and processes of dynamic self-organization will be defined according to mathematical models more adequate to describe life phenomena [11, 16, 19]. From this perspective, then, we are dealing only with two different conceptions of time, namely the time as repetition (invariance) and time as disintegration (dissolution), but we find ourselves before a third concept of time able to overcome this dualism: the time as construction, a construction that appears to our eyes, simultaneously, as creation and as rediscovery, although this same construction goes through specific states of degradation and invariance. This weft presents itself, at the same time as creation and revelation. As the creation of new forms of autonomy, and at the same time as continuing revelation of new levels of generative power: an emergence of ever new meanings that shape consecutively and in a closer manner the determinations of time, which form, in turn, on the basis of precise mathematical forms, the variegated and constrained expression of the language of life.

References

1. Anderson, P. W., & Stein, D. L. (1985). Broken symmetry, emergent properties, dissipative structures, life. Are they related? In F. E. Yates (Ed.), *Self-organizing systems: The emergence of order*. New York: Plenum Press.
2. Andrecut, M., Foster, D., Carteret, H., Kauffman, S. A., et al. (2009). Maximal information transfer and behavior diversity in random threshold networks. *Journal of Computational Biology, 16*(7), 909–916.
3. Atlan, H. (1987). Self creation of meaning. *Physica Scripta, 36*, 563–76. doi:10.1088.
4. Atlan, H. (1998). Intentional self-organization. Emergence and reduction. Towards a physical theory of intentionality. *Thesis Eleven, 52*, 5–34. doi:10.1177.
5. Atlan, H. (2000). Self-organizing networks: Weak, strong and intentional, the role of their under-determination. In A. Carsetti (Ed.), *Functional models of cognition* (pp. 127–143). Dordrecht: Kluwer.
6. Avery, J. (2003). *Information theory and evolution*. Singapore: World Scientific.
7. Bray, D. (1995). Protein molecules as computational elements in living cell. *Nature, 376*, 307–312. doi:10.1038.
8. Carsetti, A. (2000). Randomness, information and meaningful complexity: Some remarks about the emergence of biological structures. *La Nuova Critica, 36*, 47–109.
9. Carsetti, A. (Ed.). (2009). *Causality, meaningful complexity and embodied cognition*. Berlin: Springer.
10. Carsetti, A. (2010). Eigenforms, natural self-organization and morphogenesis. *La Nuova Critica, 55–56*, 75–99.
11. Carsetti, A. (2012). The emergence of meaning at the co-evolutionary level. *International Journal of Applied Mathematics and Computation, 219*, 14–23. doi:10.1016.
12. Carsetti, A. (2013). *Epistemic complexity and knowledge construction*. Berlin: Springer.

13. Di Bernardo, M. (2010). Ordine gratuito, morfogenesi autonoma e complessità semantica. *L&PS- Logic & Philosophy of Science, 8*(1), 31–82.
14. Fox Keller, E. (2000). *The century of the gene*. Cambridge, MA: Harvard University Press.
15. Fox Keller, E. (2005). *Expliquer la vie. Modèles, métaphores, et machines en biologie du développement*. Paris: Gallimard.
16. Freeman, W. J. (2008). A pseudo-equilibrium thermodynamic model of information processing in nonlinear brain dynamics. *Neural Networks, 21*, 257–265. doi:10.1016.
17. Freeman, W. J. (2011). Understanding perception through neural "codes". *IEEE Transactions on Biomedical Engineering, 58*(7), 1884–1890. doi:10.1109.
18. Gerstein, M. B., Kundaje, A., Hariharan, M., Landt, S. G., Yan, K. K., Cheng, C., Mu, X. J., et al. (2012). Architecture of the human regulatory network derived from ENCODE data. *Nature, 489*, 91–100. doi:10.1038.
19. Grossberg, S. (2000). Linking mind to brain: The mathematics of biological intelligence. *Notices of AMS, 47*, 1361–1372.
20. Grossberg, S. (2007). Towards a unified theory of neocortex: Laminar cortical circuits for vision and cognition. *Progress in Brain Research, 165*, 79–104.
21. Hintikka, J. (1970). Surface information and depth information. In J. Hintikka & P. Suppes (Eds.), *Information and inference* (pp. 298–330). Dordrecht: Reidel.
22. Kauffman, S. A. (1993). *The origins of order, self-organization and selection in evolution*. New York: Oxford University Press.
23. Kauffman, S. A. (1995). *At home in the universe: The search of the laws of self-organization and complexity*. New York: Oxford University Press.
24. Lakoff, G. (1971). On generative semantics. In D. Steinberg & L. Jakobovits (Eds.), *Semantics* (pp. 232–296). Cambridge: Cambridge University Press.
25. Lakoff, G., & Nuñez, R. (2001). *Where mathematics comes from: How the embodied mind brings mathematics into being*. New York: Basic Books.
26. Lefever, R., & Prigogine, I. (1968). Symmetry-breaking instabilities in dissipative systems. *Journal of Chemical Physics, 48*(4), 1695–1700. doi:10.1063.
27. Li, M., Wang, I. X., Li, Y., Bruzel, A., Richards, A. L., Toung, J. M., Cheung, V. G., et al. (2011). Widespread RNA and DNA sequence differences in the human transcrip-tome. *Science, 333*(6038), 53–58. doi:10.1126.
28. Longo, G., & Montévil, M. (2014). Biological order as a consequence of randomness: Anti-entropy and symmetry changes. In G. Longo & M. Montévil, *Perspectives on organisms* (pp. 215–248). Heidelberg: Springer.
29. Maynard Smith, J. (2000). The concept of information in biology. *Philosophy of Science, 67*, 177–194.
30. Mori, H., & Kuramoto, Y. (2001). *Dissipative structures and chaos*. Berlin: Springer.
31. Nicolis, G., & Prigogine, I. (1977). *Self-organization in nonequilibrium systems*. New York: Wiley.
32. Nicolis, G., & Prigogine, I. (1989). *Exploring complexity: An introduction*. New York: Freeman.
33. Prigogine, I. (1980). *From being to becoming*. San Francisco: W.H. Freeman.
34. Prigogine, I. (1988). Origin of complexity. In A. C. Fabian (Ed.), *Origins: The Darwin college lectures* (pp. 69–89). Cambridge: Cambridge University Press.
35. Prigogine, I., & Stengers, I. (1979). *La nouvelle alliance. Métamorphose de la science*. Paris: Gallimard.
36. Prigogine, I., & Stangers, I. (1984). *Order out of chaos*. New York: Bentham Books.
37. Talmy, L. (2000). *Toward a cognitive semantics*. Cambridge: MIT.
38. Waddington, C. H. (1942). Canalization of development and the inheritance of acquired characters. *Nature, 150*, 563.
39. Waddington, C. H. (Ed.). (1970). *Towards a theoretical biology*. Chicago: Aldine Publishing.
40. Weber, A., & Varela, F. (2008). Naturalizing teleology: Towards a theory of biological subjects. In L. Illetterati & F. Michelini (Eds.), *Purposiveness* (pp. 201–219). Frankfurt: Ontos.

Chapter 15
Changing Framework in Explaining Complex Dynamics: Convergences on Systemic Accounts from Two Different Case Studies

Nicola Di Stefano and Marta Bertolaso

1 Introduction

Systems that continuously acquire new, multiple, superimposed and often delocalized coherent sequences of properties show a complexity that cannot be grasped by traditional approaches that focus on fixed systemic properties. Properties in the latter approach can be grasped, for example, through explanatory concepts like self-organization and emergence, while in the former approach the history of the system has to be taken into account as well. That is, the dynamic inter-change of the systems and of their physiological environment has to be considered at once. There is, therefore, a shift in the explanatory focus that requires both a reformulation of problems and a post-reductionist trans-disciplinarity precisely related to the study of such reformulations and their properties.

In this paper we compare two case studies—music perception and cancer—by looking at how interpretative models evolve while trying to explain the dynamic features of the underling process. In middle Sects. 2 and 3 we introduce cancer case and music perception case. In Sect. 4 follows a discussion of the dimension which can be differently grasped by apparently divergent models and of the peculiar context-dependency of such properties, that will be called embodied-context-dependency. Finally, we explicit a potential outlook in term of transdisciplinarity of the analytic dimension outlined in this paper, for a deeper comprehension of the understanding process of complex dynamics in biological systems.

N. Di Stefano (✉)
Institute of Philosophy of Scientific and Technological Practice, University Campus Bio-Medico di Roma, Via Alvaro del Portillo 21, 00128 Roma, Italy
e-mail: n.distefano@unicampus.it

M. Bertolaso
Faculty of Engineering and Institute of Philosophy of Scientific and Technological Practice, University Campus Bio-Medico di Roma, Via Alvaro del Portillo 21, 00128 Roma, Italy
e-mail: m.bertolaso@unicampus.it

G. Minati et al. (eds.), *Towards a Post-Bertalanffy Systemics*, Contemporary Systems Thinking, DOI 10.1007/978-3-319-24391-7_15

2 Cancer Explanatory Theories

2.1 The Discrete: Gene Centered Models

Some cancer explanatory models tend to focus on the cellular component, whose abnormal behavior is due to genetic and epigenetic factors (we will refer to them as "genetic centered perspectives" or GCP). In GCP, cells mainly act as autonomous units that can eventually fully account for cancer. GCP explains cancer in terms of a sequence of mutations of genes, or epigenetic alteration that mimic genetic mutation transforming normal cells in cancer cells. To a first approximation, tumors' properties are explained by the properties entailed in the cell and its components. Tumor cells' properties are traditionally described as follow [7]: self-sufficiency in growth signals; insensitivity to antigrowth signals; evasion of apoptosis; limitless replicative potential; sustained angiogenesis; invasion and metastasis. These hallmarks of cancer focus on the autonomy of tumor cells from other cells and from their environment. They might characterize all tumor cells. Yet this is not the case. Tumor cells are not completely autonomous, and tumors' cells heterogeneity leaves open the question about the explanatory role of cancer cells characterized in those terms. Even attributing to cancer cells the capability to develop all features of a neoplastic phenotype, GCP has difficulties in accounting for tumor heterogeneity. By definition parts should be homogeneous but, in fact, they are not.

2.2 The Continuous: Cell Centered Models

In the last 50 years, however, the general emphasis has changed more and more from considering cancer as the progressive result of genetic mutations producing defects in the regulatory circuits, that regulate proliferation of normal cells, to considering cancer as "a disease of cell differentiation rather than multiplication" [8]. The discovery of Tumor Suppressor Genes (TSGs) integrated the Somatic Mutation Theories universally agreed-upon model of cancer origin and progression. In those years, experimental observations called attention to a different dimension of carcinogenesis other than cell proliferation. The fact that normal mouse cells showed a dominant behavior over cancer cells when the two cell types were merged [9], suggested that the former had to carry genes that were opposed to tumourgenesis, thus showing a tumor-suppressor function. This and other data suggested that carcinogenesis does not require an acquisition of function (assumption that was behind the definition of oncogenes), but rather the loss of it, perhaps through damage of some pattern of cellular differentiation. This step, which is conceptually very simple, technically called for a major research effort to demonstrate that the dominance of oncogenes was not the general rule, casting the foundations of subsequent theories of the cancer cell. From an explanatory point of view, the Clonal Genetic Model of Cancer integrated the knowledge we had till then on oncogenes (ONGs) and tumor suppressor genes (TSGs) within a unified picture. At its simplest, the model predicts the functional impairment of these genes as the origin of a tumor cell's development, by clonal reproduction into the cells that constitute the tumor mass and ultimately trigger metastases.

It was Bert Vogelstein who shed light on the role of ONG and TSG in describing how different mutations were related and connected in the origin and progression of tumors. The idea of the clonal evolution of tumor cells was reinforced through the description of a coherent and sequential tumorigenic process, consisting of a series of stages of molecular events. This Multigenic and Multiphasic Model of Cancer [5] was widely accepted by the scientific community and provided the scientific basis for the initiation, promotion, transformation and cancer progression, which until then were based on purely theoretical hypothesis. What's new in this model is the emphasis on the fact that the total accumulation of molecular changes were more important for tumor progression than their sequence or even identity, concluding that five or more genetic alterations were probably required for the development of carcinomas, while a lesser number of mutations would be required for benign tumorgenesis [16]. Pointing to this much more relevant aspect of carcinogenesis, Vogelstein had the merit to conceptualize tumor genes in a different and epistemologically relevant way with respect to the kind of classifications that had been adopted since then.

2.3 The Context Dependency: Tissue Centered Models

Organism Centered Perspective (OCP) sees cancer as a problem of tissue organization so that cells lose their contextual control. From 1999 [13], some authors developed the idea that cancer has to be addressed in terms of tissue organization. Interactions between stroma and epithelium when compromised by the action of hormones or other chemical or physical factors become crucial in carcinogenesis. Hence, the study of complex phenomena, such as cancer, can be better analyzed from the perspective of higher hierarchical levels of organization than from the pure cellular level. Epistemological presuppositions assume the hierarchical organization of a multicellular organism system as default. Cancer is thus viewed as a problem of development, an organogenesis that goes awry. It is studied in terms of the three-dimensional structure of tissues and of how intercellular communication takes place within them. The cellular components, once integrated into the tissue, are meant to interact in a way that presents a new and unique feature of intrinsic reciprocity of relationship among cells in the tissue. Heterogeneity is better understood here, but anyway these models have difficulties to grasp the epistemological meaning of the multilevel phenomenology of cancer, i.e. the causal relevance of different levels of the biological organization. OCP reflects organizational systems (i.e. non-aggregative systems like in the antireductionist philosophical accounts). When we artificially separate the components of the tissue cells that form the epithelium and the underlying stroma, they stop carrying out the functions performed when they were assembled in their unique three-dimensional organization. Once recombined, instead, they form a tissue similar to that of their origin [14].

3 Musical Consonance and Dissonance

We now move to the second case study by briefly introducing a particular aspect of music perception (for consonance and dissonance case see also Di Stefano and Bertolaso [4]). Consonance and dissonance are musical notions, which have been widely studied from the ancient Greek philosophy to nowadays. Without going any further in their historical evolution, we can consider consonance as the relation that links two or more sounds sounding pleasant together, while dissonance as the relation that links two or more sounds sounding unpleasant or rough together [15, 17]. From arithmetic to physics, from psychology to physiology, different explanation of the same phenomenon have been given.

3.1 The Discrete: Frequency Ratio of Intervals

Pythagoras, in the sixth century BC, discovered that the simpler the frequency ratio, the more consonant the interval: the most consonant interval is the octave, which is expressed by frequency ratio of 2:1; frequency ratio of 3:2 gives fifths and frequency ratio of 4:3 gives fourths and so on decreasing in consonance. Thus, the Pythagorean "tetractys", composed by 1-2-3-4, gives reason for all the perfect consonances [3]. Since Pythagoras' mathematics, far from being empirical, is metaphysic, numbers are the deepest reasons of reality. That's the reason why the frequency ratio has so fundamental importance in consonance and dissonance phenomena:

> The issue of consonance and dissonance was for the Pythagoreans not a matter of devising a theory that was harmonious with their hearing, but rather one of hearing the numerical truth that they discovered to be inherent to nature [1, p. 216].

The arithmetical model considers sounds in abstract, and thus enlighten the discrete dimension of the phenomenon. If the *explanandum* is consonance and dissonance, integers ratios are the *explanans*.

3.2 The Continuous: Waves Matching

Frequency ratio is anything else than the arithmetical expression of how the two sound waves match. Consider, for example, an octave: frequency 2 is frequency 1 doubled. So, in physical terms, this means that every two peaks of the upper the two tones beat together. Since acoustics considers complex sounds, composed by several harmonics, a single tone results from different partial waves matching in different ways. Therefore, the way they match is crucial for the perceptual effect on listener [11, 12]. This represents an essential perspective's change, which leads to the notion of sounds as compound. Physics emerges as a new explanatory level able to capture the continuous dimension of consonance and dissonance perception. In such account, a sound is no more adequately represented by frequency ratios, but better represented by a continuous entity as a wave. However, as arithmetic does, neither physic pays much attention to the physiological/psychological structures of the perceiving body.

3.3 The Relevance of the Context: Regularity of Beating and Cochlea's Properties

Since consonance and dissonance are always perceived by human being, there is the need to shift to the body and to its structures. This can be considered a physiological approach, in which human component (anatomy of hearing apparatus) and physical wave properties are considered together from a unitary perspective. In this perspective, consonant intervals are associated to most regular beating of tympanic membrane. Regular means more synchronized. Helmholtz discovered that the elasticity of the basilar membrane in the cochlea is non-homogeneous, so that sounds at different frequencies have different effects on the perceiving human system [6, 10]. At this level, the physiological one, dynamic features become fundamental and therefore context starts to be relevant. If consonance and dissonance phenomenon is the *explanandum*, now embodied dynamics are the *explanans* [4]. This shift implies a deep change in perspective that highlights a fundamental property of music perception (as happened above for cancer pathology): its context dependency (see Table 1). Such context dependency lies on the fact that living structures are not homogenous neither discrete. In arithmetical terms, all fifths C-G are identical because their frequency ratio is always 2:3. But when listening to consonance, arithmetic no more describes the phenomenon adequately: hearing a fifth at 16 Hz is really different from hearing the same interval at 16 kHz, because the basilar membrane response to the same interval changes at different frequencies.

4 Analytical Dimension of the Phenomenon: The Relevance of the Context

It is, therefore, increasingly clear that the understanding of the processes evoked in the two case studies exceeds a mere psychological upshot or acquisition of knowledge of facts although a kind of relationship between understanding, explanation and contextual factors have to be hold. In particular, the focus of all these issues remains the question about how such relationship should be understood in explanatory and in conceptual terms [4].

When the context becomes relevant the relationship between the system and environment start to be semantic, and their dynamic interaction become essential. At discrete and continuous level yet no interaction is necessary. Semantic, here, means that something became significant in different ways as it differently interacts with the context. Clear examples of such dependency are offered by the OCP we referred to above. For example, when skin cells are placed in a culture dish, they form a uniform layer of tissue very different from the original, although, if they are placed on a surface previously covered with basement membrane proteins, they tend to associate among themselves and recover the original three-dimensional structure of the epithelium from which they came. However, they often undergo genetic transformation and imbalance when cultured for a long time in vitro so that the real future challenge can be nicely summarized in a title like this: 'Decoding the language of form' [2].

Similarly, in music, the same orchestral excerpts played by different instruments (i.e. within different context) would give totally different result in terms of perception. In arithmetical terms, the original version and the altered one are exactly the same (same frequency ratio). From the physics perspective, they are different because sounds' matching is different, and corresponding waves are different. From the physiological/psychological perspective, they are different not only because sounds matching is different but also because human perception reacts differently.

EXPLANATORY LEVEL	ANALYTICAL DIMENSION	SYSTEM-ENVIRONMENT
ARITHMETIC	*Discrete*	SYSTEM
PHYSICS	*Continuous*	
PHYSIOLOGY/PSYCHOLOGY	*Context dependency*	SYSTEM-ENVIRONMENT
GENE	*Discrete*	SYSTEM
CELL	*Continuous*	
TISSUES	*Context dependency*	SYSTEM-ENVIRONMENT

Fig. 15.1: Relationship between explanatory accounts and analytical dimension in the studied cases (cancer and musical consonance and dissonance). Modified version of Table 1 in [4]

The relevance of the context and its properties is evident, for example, in Baroque music, where we often assist to the introduction of a dissonance creating a sort of a new coherence which guides the perceptual and cognitive activity of the listener.

The emergent character of phenomena considered appears, in epistemological terms, when the context dependency becomes relevant (see Fig. 15.1). At the same time, context dependent levels of explanations are more comprehensive than the discrete and continuous ones and such emergent character might have some explanatory priority over the phenomena considered. In fact, phenomena can be distinguished and not separated in all accounts, though it is in the context dependent level that their inter-relationship and their unitary outcome (whether cancer or music perception) actually becomes the very object of inquiry. In this sense, systemic perspective and trans-disciplinarity, as previously explained, become necessary and unavoidable.

Therefore, we cannot definitely choose which is the privileged level the phenomenon should be explained at: there is no chance for any *experimentum crucis*, neither in cancer nor in consonance and dissonance perception, because there is no unique privileged causal level. Systemic approach widen the perspective to the living context, where arithmetic, physics, biology and physiology are fused in a complex and dynamic process. Every attempt to find the definite reason or cause only in one of this partial aspect loses the unity of the whole. Moreover, as happens for more complex biological dynamic behaviors, systemic approach avoids the risk of making methodological recommendations about the ontological restrictions—frequent in epistemological or ontological reductionism—and also avoids deriving methodological indications from holistic principles.

5 Conclusions: Levels of Understanding Process

Facing problems like cancer pathogenesis or musical consonance and dissonance we have to manage a complexity growing up from the biological ground. Correctly understanding what the problem is, or at least what the problem is not, implies, in some cases, a shift within the same discipline, like in cancer case, and in other cases a shift between different disciplines. In the first case we need intradisciplinary shift, in the latter, we need interdisciplinarity. Every authentic widening of perspective really represents an advantage in understanding, though the result of the match between different disciplines can be very different. Multidisciplinarity is the way several disciplines study the same problem, each one maintaining its peculiar basilar concepts. The results are in this case merely combined and their sum is weak. Interdisciplinarity happens when disciplines learn from each other's approaches, concepts, theories, methods. Though, in this case, there is real dialogue between disciplines, it lacks the creation of a new point of view.

Transdisciplinarity is the creation of a new theoretical framework, potentially opened to the emergence of a new discipline. This is the strongest approach and the only one that really leads to a new phase in problem understanding, which may bring to the redefinition of the object of inquiry in more adequate terms. Transdisciplinarity manages every account as a whole-dependent level and not as something concluded in itself: approaching multidimensional phenomena, accuracy of an explanatory account depends not only on the level of details gained through different methodologies, but also on interplay and reciprocal dependence between the scientific question and the phenomenon under inquiry. Transdisciplinarity, bringing back into the same inquiry the system and its context, allows to distinguish and clarify the epistemological relevance of what we have defined analytical dimensions of phenomena. It also explains how the emergence of what is explanatory relevant structures the process of understanding embodied dynamics. A systemic perspective, therefore, allows us to avoid excessive simplifications driven by mere pluralistic accounts of human understanding and scientific knowledge. Depending on the explanatory contexts, meanings can change and require different epistemologies. The novelty of a systemic and transdisciplinary approach lies not on its object but on the way the object is approached.

Contextual factors in the process of identifying the explanatory level and the relational principles of organization structure the process of understanding. The discovery of new explanatory relationships characterizes such process, i.e. different kinds of understanding emerge as different levels of comprehension of explanatory relevant relationships.

Therefore, the irreducibility of understanding is not an obstacle but rather a condition of the integration of different kinds of human understanding. What might eventually emerge is an epistemology able to ground real interdisciplinary or—more precisely—transdisciplinary approaches which overcome the epistemological tensions on causal and explanatory notions still affecting the philosophical consideration on the understanding of multi-level and systemic dynamics.

References

1. Barbera, A. (1984). The consonant eleventh and the expansion of the musical tetractys: A study of ancient Pythagoreanism. *Journal of Musical Theory, 28*(2), 191–223.
2. Bissell, M. J., Kenny, P. A., & Radisky, D. C. (2005). Microenvironmental regulators of tissue structure and function also regulate tumor induction and progression: The role of extracellular matrix and its degrading enzymes. *Cold Spring Harbor Symposia on Quantitative Biology, 70*, 343–356.
3. Crocker, R. L. (1963). Pythagorean mathematics and music. *The Journal of Aesthetics and Art Criticism, 22*(2), 189–198.
4. Di Stefano, N., & Bertolaso, M. (2014). Understanding musical consonance and dissonance: Epistemological considerations from a systemic perspective. *Systems, 2*(4), 566–575.
5. Fearon, E. R., & Vogelstein, B. (1990). Genetic model for colorectal tumorigenesis. *Cell, 61*, 759–767.
6. Frosch, R. (2012). *Musical consonance and cochlear mechanics*. Zurich, Switzerland: VDF.
7. Hanahan, D., & Weinberg, R. A. (2000). The hallmarks of cancer. *Cell, 100*, 57–70.
8. Harris, H. (2004). Tumour suppression: Putting on the brakes. *Nature, 427*, 201.
9. Harris, M. (1971). Cell fusion and the analysis of malignancy. *Proceedings of the Royal Society of London - Series B: Biological Sciences, 179*, 1–20.

10. Helmholtz, H. (1954). *On the sensations of tone*. New York: Dover.
11. Kameoka, A., & Kuriyagawa, M. (1969). Consonance theory part II: Consonance of complex tones and its calculation method. *The Journal of the Acoustical Society of America, 45*(6), 1460–1469.
12. Plomp, R., & Levelt, W. J. M. (1965). Tonal consonance and critical bandwidth. *The Journal of the Acoustical Society of America, 38*(8), 548–560.
13. Sonnenschein, C., & Soto, A. M. (1999). *The society of cells: Cancer and control of cell proliferation*. New York: Springer.
14. Soto, A. M., Maffini, M. V., & Sonnenschein, C. (2008). Neoplasia as development gone awry: The role of endocrine disruptors. *International Journal of Andrology, 31*, 288–293.
15. Trainor, L. J., Tsang, C. D., & Cheung, V. H. W. (2002). Preference for sensory consonance in 2- and 4-month-old infants. *Music Perception, 20*(2), 187–194.
16. Vogelstein, B., & Kinzler, K. W. (2004). Cancer genes and the pathways they control. *Nature Medicine, 10*, 789–799.
17. Zentner, M. R., & Kagan, J. (1998). Infants' perception of consonance and dissonance in music. *Infant Behavior and Development, 21*(3), 483–492.

Chapter 16
Perceptions of Landscape: Observed and Observing Systems

Valerio Di Battista

1 Introduction

The "perception" of landscape allows a more complete concept of what we name as "land" or "environment". The definition of "territory" still inherits the modern interpretation that favors the security needs (vulnerability to natural events likely to be dangerous) and the functions of usability (potential settlements, land use, presence of resources, infrastructure, services, etc.). The definition of environment refers to all the complexity of geographical and ecological conditions and to the definition of requirements for human comfort and health.

The idea of landscape "as perceived by the people" integrates the above concepts with the system of signs and symbolic meanings that is essential to our subjective and socio-cultural identity. This consideration, absent or underestimated in the usual concepts of territory and environment, when denied, leads to "disorientation". In extreme cases it can provoke mental illness, and it is even used as a torture technique. The landscape—made of signs and meanings of territories and environments) increasingly represents the metaphor of our relationship with the world, with others, with ourselves. For each population, the "right" landscape represents signs and meanings of the settlement system (territory, environment and culture) where one belongs. For each individual it encompasses the symbols of relations with spaces, people, things with which one lives. Every cultural landscape always represents the match between the natural configuration of a place and the actions (intentional and unintentional) exerted by the settled population for their material and cultural utility. For each population, the "right" landscape encompasses the signs and meanings of the settlement system of belonging. They also represent

V. Di Battista (✉)
Politecnico di Milano, Milano, Italy

Rete Piemontese degli Osservatori del Paesaggio – Piedmont Network of Landscape Observatories, Provincia di Alessandria, Italy
e-mail: dibattista_valerio@libero.it

G. Minati et al. (eds.), *Towards a Post-Bertalanffy Systemics*, Contemporary Systems Thinking, DOI 10.1007/978-3-319-24391-7_16

the output of both intentional projects, and the interaction of many unintentionally connected actions and goals, collective, unintentional, even accidental that change the landscape during time.

"Unintentional project" is what we call this summation of chaotic and ongoing actions. This evokes the idea of possible forms of self-regulation. Along these lines, I developed the following parallel and has dealt with two reflections: the first one deals with the CEP definition of "landscape as perceived by the population"; the second one refers to the "field test" launched in Monferrato in 2006 [3] and currently going on in the experience of "monferratopaesaggi.org".[1]

2 Difficulty in the Interpretation of Landscape

Theoretical frameworks on the European CEP mainly concern the relations between the different perceptions of landscape by either "the experts" or by population. In the "perception of the landscape" these different perceptions regard the interaction between observed and observer systems, thus posing questions of systemic nature. The various expert disciplines studying and describing the landscape—geography, environmental sciences, planning, social and anthropological studies and the such—represent one end of the observing system. They seem not being able to keep in touch with the other end: that is, with all those entities who actually operate in the observed system itself (people, institutions, etc.). Populations who in a more direct way perceive landscapes through senses and emotions, seem to be scarcely aware of the achievements expert studies bring forth. Both observing groups (populations and experts) usually perceive one another as bearers of abstract interpretations. Failed dialogue between the different interests of the observers results in chaotic interaction of the various processes. These unresolved separations explain the difficulties of interpretation of the definition of landscape introduced by the CEP. These difficulties arise in matching the contents of the "landscape as perceived" by people, with the tools of the cultured observers, the learned disciplines. Bottom-up activities promoted by the populations do not fit with top-down activities, promoted by the various institutions (from above). To overcome these difficulties we need to consider the perception of the landscape as a way of reading the relationship between the "things", the signs, and the people who enjoy them and "manage" their meanings. The "things" of the landscape express meanings that the observers give them, but as the observers themselves act in the landscape they are also an inseparable part of the observed system. They literally are part of the observed system and must be considered along with the tangible and intangible elements of the landscape itself. But as observer systems structure their observations according to the peculiarities of the observed systems, the concept of landscape undergoes a perception based on the unstable relationships between observers and observed systems that generate fields

[1] www.monferratopaesaggi.org is a portal created by the Observatory of the Landscape for the Monferrato and co-funded by the Compagnia di San Paolo.

of variability hardly operable. This relationship develops between the places of life, both in their practical and symbolic aspects, and those who build and deconstruct them, materially and immaterially, by operating inside them. In these processes, an uneasy relationship arises between intentionality, intrinsic to all knowledge, and all those unintentional actions arising from numerous and unconnected actions operated by all landscape stakeholders. All this explains the diversity and legitimacy of the different perceptions of the landscape and highlights all the difficulties of interpretation, analysis, and management of preservation/development projects.

We have analyzed these issues and work in the activities of the landscape for the Monferrato, developed in a hilly area of 523 km^2, which includes 38 municipalities (9 in core areas Unesco) and a population of about 65,000 inhabitants. The description of the landscapes of Monferrato began over 10 years ago[2] and highlights this language problem. Our language, however, rich in adjectives and shades, struggles to describe, for example, the differences between urban and rural landscapes which are apparently similar. A similar and more surprising inadequacy is evident in the reproduction of images where the many possible techniques (photos, videos, drawings, etc.), find it difficult to give back the direct "perception" (visual, sensory, psychological, cultural) that we experience only with our presence "on the spot".

These difficulties in describing the "perception" of landscape show how this problem is still very often confined to a realm which is unspeakable, totally emotional. This fact underlines the difficulty of comparing interpretations and values. Thus we decided to research descriptive parameters which could be measured, compared and used as references for preliminary experimental evaluations. These can be, at first, reductive but as far as possible independent of subjective interference.

3 The Case Study

The Observatory, a result of a research programme co-funded by the Compagnia di San Paolo, has been able to implement the project *Monferrato Paesaggi*. We assumed an "expert assessment of visibility" as a necessary starting point which could allow to later apply some techniques of direct listening. We directed expert observers to assess a few landscapes characters considering practical possibilities of visibility (paths, points, conditions) offered by the area. These have been determined primarily by the possibility of being perceived from panoramic points of view along the roads selected both for features such as openness, depth and character, and such as the number of users, time and frequency of use. Different portions of the landscape can be assessed comparing various types of tracks and probabilistically according to the terms and conditions of use. This suggested to choose the paths most likely to be used, in this case those with low intensity traffic and high visibility. Measurable and comparable assessment criteria adopted were: continuity and length of routes with panoramic opening; breadth and depth of the visual fields; the variety and consistency of character. In addition, we concepts such as: active

[2] See www.odpm.it.

visibility (the field of view perceived by the observer system) and passive visibility (the perceived visual field of the observed system); the dominant view (emergencies characterizing the territory); the prevalent configurational characters and, both locally and in general, the number of points of view and places of interest.

The above parameters represent the boundaries of the observation field. Along two sides run the geographically dominant edges: the Po River to the north, the hill foot margin with the plain to the east. To identify the other two sides, without equally obvious elements, we considered the "boundaries" most recognized for social belonging. The main geographical elements: the Po plain and the Alps to the North; the sequence of ridges that evokes Renaissance landscape paintings to the South, are the dominant images that define the identity of the landscape of the entire area. Further on, four sub-areas with different landscape features have been identified.

These general indications define the identity characteristics of the territory: that is, those that distinguish it from others with similar configurations, thus providing a first "test" on paths, wide area, and historic small towns which were conducted along with all the interested municipalities. These tests have been developed along with the explorations conducted to distinguish the territories with different characters and select the paths that could be significant for the perception of the landscapes and their visual relationships. With the same criteria, but with more focus on the more complex and detailed characters (road structures, building types, attractions, monuments and artifacts of material culture) we considered the historical small towns.

The urban landscape of Casale Monferrato, the most important medium-sized city in the area, with its attractive and compact historic center, posed different problems. In this case the relationship with the public administration focused around the drafting of a map of selected paths: either to provide urban images are representative of the most significant local character and identity, and to connect the more relevant sites, monuments, buildings and cultural heritage. As before, we had two aims: to raise awareness among residents about the value of their urban landscape and to offer interesting routes to outside visitors. Here, the interaction between observers (subject experts) and observed landscape was conducted with the study of perceptual and descriptive view in different areas and on different paths. We tested various methods and procedures to analyze the "local cognitive perception", determined by the common culture and local uses giving meaning and value to places and things. A cognitive "perception", collected by sampling heterogeneous emotional impressions, compared and reconstituted at the level of groups and communities, seems to get close to represent the social perception evoked by CEP. Among the methods we used significant views and items as selected by "experts" to identify possible social perceptions exercised by residents or visitors. In general, in the campaign of observation, surveys were submitted to administrators and to many population samples in order to gain information on perceived values to compare them to the experts' perceptions. These surveys were useful to generate sensitivity and interest but they also showed an unexpected and widespread lack of knowledge of the territories. Interest resulted mostly motivated, when present, by practical reasons, such work engagements or relatives living in places other than those of residence. Main attractions resulted to be the bigger cities.

Residents expressed a general appreciation for the landscape, though a low awareness of its values and potentials. Visitors and temporary residents owners of holiday houses show higher levels of satisfaction and value assignment. The somewhat low appreciation of the landscape by local residents mostly shows certain factors of addiction and inattention to the environment, but is accompanied by a lack of knowledge of the area and by a widespread difficulty of understanding places, even in the face of impressive sights. An interesting element is that the appreciation of centers and historic buildings has spread only when new denotations of rarity and interest have given more value to environments, even when quite humble and characterized by memories of poverty and fatigue (farms, peasants' cottages, former factory buildings). It is likely that even for some landscapes, negative evaluations come from such memories (hard work, reduction of population, lack of services) generating neglect. The lack of knowledge of places and landscapes, ignorance of their qualities and values all depend from systems that use cognitive filters inadequate (cultural-language models) or incongruous (comparisons with other sites and landscapes of tourism consumption). But it also depends upon the conditions of degradation and neglect in the observed systems. These considerations seem to require procedures that increase knowledge, recognition of values, promotion, protection procedures and retraining.

4 Conclusions

The process started on the Monferrato landscape aims to increase knowledge and appreciation of the places, thus stimulating local maintenance and management processes. The landscape assessment program, conducted with the participation of local governments and associations, have selected and proposed six routes (335 km in total, of which 101 with panoramic openings), and 12 access paths with high landscape interest, 35 routes in historic towns, 29 interconnected pedestrian paths and 5 thematic routes and reported more than 500 points of interest and 250 points of view. The detailed survey work increased interest and knowledge in the local population, led to organize paths networks, highlighted places and points of interest, adding value to the area assessed. It is now up to the various actors (systems observers and decision makers) to embrace this procedure and start promoting the entire network. Next step will be the development of a network of interactive indicators (posters, totems, QR codes and signs on the ground), along with a dedicated interactive website (www.monferratopaesaggi.org) and a program of educational meetings in all communities. The website is georeferenced and can be used on computers, smartphones, tablets.

Items are: Land, Landscape, routes, municipalities, about us. The item "Territory" contains general descriptions of: History, Geology, Population, Vegetation and flora, fauna. The item "Landscape" explicit methods and outcomes general readings conducted. The item "Routes" gives access to the entire network of existing routes (car, bicycle and pedestrian, "Municipalities" provides for each zone quick descriptions and images, describing the itineraries of historic.

The entire project follows the principle that landscape becomes a resource if it is accepted, it is accepted if it is known and appreciated, and it is appreciated if it is cared for, maintained and improved. Therefore, the website is also addressed to visitors, but it is most of all designed to accept comments and generate attention by all local inhabitants and business. These are the main actors of what we call an "unintentional project"; in our understanding, the use of this site should promote communication, transparency, awareness and attention to all changes that will occur in the area.

References

1. Badiou, A. (2006). *Il secolo*. Milano: Feltrinelli.
2. De Luzenberger, G. (2004). *Breve guida all'uso della metodologia European Awareness Scenario Workshop*. Firenze: Scuola Superiore di Facilitazione.
3. Di Battista, V., Giallocosta, G., & Minati, G. (2006). *Architettura e Approccio Sistemico*. Milano: Polimetrica.
4. Luginbühl, Y. (2012). *La Mise en scène du monde. Construction du paysage européen*. Paris: CNRS Éditions.
5. Luginbühl, Y. (2013). Conclusion. In Y. Luginbühl & D. Terrasson (Eds.), *Paysage et développement durable*. Versailles: Quæ.
6. Osservatorio del Paesaggio per il Monferrato Casalese (a cura di). (2012). *Sistema paesaggio*. Milano: PoliPrint.
7. Sempieri, A. (2008). *Nel paesaggio*. Roma: Donzelli.

Chapter 17
Thinking Smart City with a Focus on Emerging Identity Elements

Valentina Volpi, Antonio Opromolla, Roberta Grimaldi, Valerio De Cecio, Mauro Palatucci, and Carlo Maria Medaglia

1 Exploration and Identification of Urban Codes and Languages Emerged by Collective Actions

The city is a system continuously changing and evolving. It modifies it-self on the basis of patterns that are in fact languages and codes commonly shared by people.

Generally, the language has basic elements that are combined in different ways, generating in so doing different structures. On the other hand, the language is subject to change in respect of the use made of it.

In the city, the language elements are both architectural and social. They may follow fixed patterns, but all the variables occurred in the architectural design and urban planning are not predictable. Moreover, the evolution of the city over time may imply an abandonment of the elements that cannot be integrated in the contemporary context. Sometimes it simply implies a different use of these elements.

This process appears evident if we consider the city as a system affected by the interactions occurred within its boundaries. Moreover, referring to a systemic approach, the concept of "city" is strongly bound with that of the "emergence". According to Minati [6],

V. Volpi (✉) · A. Opromolla · R. Grimaldi · V. De Cecio
ISIA Roma Design, Piazza della Maddalena 53, 00196 Rome, Italy

Link Campus University, Via Nomentana 335, 00162 Rome, Italy
e-mail: valentina.volpi84@gmail.com; anto.opro@gmail.com; robertagrim@gmail.com; valerio.dececio@gmail.com

M. Palatucci
ISIA Roma Design, Piazza della Maddalena 53, 00196 Rome, Italy
e-mail: mauro.palatucci@gmail.com

C.M. Medaglia
DIAG, Sapienza Università di Roma, Piazzale Aldo Moro 5, 00185 Rome, Italy
e-mail: carlomaria.medaglia@uniroma1.it

G. Minati et al. (eds.), *Towards a Post-Bertalanffy Systemics*, Contemporary Systems Thinking, DOI 10.1007/978-3-319-24391-7_17

> emergence deals with a generalization of processes of self-organization by considering the process of hierarchically acquiring new properties as properties of systems of systems. Through processes of emergence systems acquire themselves or collectively (i.e., through systems of systems) new further systemic properties at different levels.

In fact the emergent properties, included those affecting the city, are not predictable.

Despite this indefiniteness, the emergence of new properties in the "city" system corresponds to the development of the territory and the reaffirmation of its identity.

In such a context, the human factor is able to continuously generate new unpredictable attributes, affecting the essence of the identity of a city. So, we consider as useful reference for our work the

> "Human Smart City" approach, "where people – citizens and communities – are the main actors of urban 'smartness'. A Human Smart City adopts services that are born from people's real needs and have been co-designed through interactive, dialogic, and collaborative processes" [5].

The complexity of the factors implied in this process of city shaping made necessary the adoption of systemic to study the possible forms that a city may assume during the evolution process.

Indeed, according to Minati [6]

> human settlements are the product of human societies, they are mostly built and developed by a huge number of unconsciously interacting acts over a long time, rather than by purposely designed single acts [6].

Without considering the complex interdependence among the numerous interactions and occurrences arising in a city, that express its vital identity, all the actions operated in order to increase the quality of life of the city users risk to be rejected by the system.

The aim of this paper is to identify and emphasize on processes of emergence that may help designers thinking about possible targeted design solutions to appropriately consider the continuously changing state of a city. Moreover we research about elements that affect the identity of a city, whose essence is continuously redefined by the emergent properties sprung from the actions and interventions of people. So three different examples are illustrated to sustain our thesis: the re-semantization of specific city elements, the boundary conditions as source of system change, and the unconventional social behaviours as lurking soft protests.

2 Re-semantization of Specific City Elements

The re-semantization in linguistic context consists in the attribution of new meaning to an existing lexical element. The increasing complexity and the multiple articulation that characterize the city framework provide a lot of examples of attribution of a new meaning to urban ecosystem elements.

In semiotic studies, the space in which people live is presented as the expressive level of a real language. It is related not only to places but also to the communities and their self-organization, on the basis of what they consider inside and outside them.

All the different kinds of individual (human, animal, object, which are natural or cultural) inhabit and transform the space, polarizing and articulating it.

The semiotics of urban space looks at the city and its urban articulation as elements related to a linguistic system endowed with meaning.

This complex system of signs is presented as a set of layered elements driven by their own dialectic; the latter could be at the same time peaceful and conflictual.

For example, the road signs system represents a system of signs that overlaps that formed by the grid of streets and squares of the city, providing to people a code that describes both the legally and the illegally use of streets and squares. In reflecting on the relationship between the meaning and the code, there is often a "peaceful reading", but sometimes there is an "extremely conflicting one", as it happens when the same rules system is applied to different kinds of vehicle (e.g., car, bicycle, pedestrian, etc.) [2, 7].

In our opinion, the re-semantization can be intended as one of the expressions of the emerging properties of the identity of a city. It arises as the result of the wear process of the relationships between subjects and objects that mould the city. We do not intend the wear in its original meaning, i.e. in the negative sense of "decline as a result of prolonged use". The wear of the relationships between subjects and objects in the city consists in the innate process of changing of complex and open systems, whose survival lies in their own dynamism and in the consequent ability to manage and acquire new emergent properties. The city regarded as the intersection of processes, dynamics of use, needs, and requirements is a system in constant changing, whose identity is transformed by the people who inhabit it and by the relationships that people establish with objects.

The social action, as well as the individual one, is in a reflexive relationship with the environment and its components. It also constructs new meanings, redefines and re-establishes relationships, or modifies the old ones; namely it generates a re-semantization of the environment and the objects that compose it. An example of this process of signification is the use of elements of the urban furniture for other functions than the ones they were designed for. Most of the time, this dynamic can be represented like a "rash" of people needs unmet by the Institutions, which follows a parameter of effectiveness and efficiency in order to look for their satisfaction in an alternative way. We assume that for people it is easier to re-think something that already exists rather than thinking about something *ex-novo*.

3 Boundary Conditions as Source of System Change

The new identity elements of the urban system can be also related to the "boundary conditions".

In our opinion, a city can be compared to a set of concentric circles in which the external ones (the boundary area) can be considered as the emerging properties of the system, and the internal ones (the central area) as the old systemic properties.

This consideration is mainly related to the spatial conditions; for example, the city centre contains the most important institutional, architectural and political elements of the city, whereas the suburbs contain the marginal cultures. However, this consideration is not limited to a spatial representation. For example, it can be also related to the topics of the political and media agendas: in them, some themes are more important than others, and this hierarchy is often related to the history and traditions. Moreover, the opposition between the emerging properties of a system and the old ones, it is not only a cultural factor (leading culture vs countercultures or emerging cultures), but it is in general related to individual and specific attributes of the identity.

In this context, the emerging properties can be considered as transformation elements of the system that modify, shape and reorganize its structure and its conditions. From this point of view, we can affirm that the "boundary" is the source of the system change, whereas the "centre" is the core of the system identity. In a territorial system, the tendency to preserve the old properties can be very strong and the boundary is considered also the vehicle employed in order to reaffirm them: indeed, it points out the difference between what is on this side and what is beyond the boundary [3].

Considering that, the boundary is both a barrier (so a disabling element for the identity) and a bridge (so an enabling element) [4]. In the latter case, the new proprieties of the boundary reach the centre, originating fusion, contamination, innovation, and opportunities. This process is the outcome of individual or collective actions that aim to add these new properties to the system identity. It can be a process more or less "dramatic" and "extreme". It mainly depends from the distance of the new properties from the old ones: if this distance is very large, the acquisition of the new identity proprieties is a proper "revolution". Moreover, specific events can accelerate this process; on the contrary, if these events do not occur, the acquisition process can be more peaceful.

In any case, the aim of these processes is to lead the system (the city) to a more realistic identity, by ensuring that all the social, political, and cultural phenomena have an equal consideration.

An example of these observations are the French riots occurred in 2005 in the *banlieu*, when the discontent of the Arab, North African, and black French second-generation immigrants suddenly exploded, due to the low consideration of their needs in the French policies.

In most cases, included the given example, the opposition between the "imposed" identity and the lived and real identity causes a proper destruction and a real break up. It is a necessary action, in order to reformulate the identity and to shape a new one.

4 Unconventional Social Behaviors as Lurking Soft Protests

Another type of emerging element that can affect the city aspect is directly connected to the social behavior and the relationship among the city users. In details, some collective actions emerge as social behaviors indicating a dysfunction of the city. These actions cannot be foreseen, but they quickly spread among people with the aim of destabilizing the system. The large distribution and the repeatability of these advocated behaviors are functional to the acquisition of the new properties, since they are generally looking for reaching a critical mass, in order to obtain a substantial reaction from the system. These actions represent an alternative solution for living the city and the services and resources that it offers. They stress some practice at the boundary of legal (without manifestly fall into the illegal) in order to complain for a more sustainable way of life. It represents a collective intelligence applied to the city to resolve common criticalities, assuming the form of a lurking soft protest. What emerges is the need to improve the life of city users, and the quickly spread of the soft protest shows the systemic nature of the need itself that affects the whole idea of city.

Some of these collective actions redefine the social interaction on the basis of mutual support and sharing "philosophy". For example, as a replay to the increase in the ticket price, when getting of by public transport, such as public bus, people start to give their still valid ticket to other passengers catching the public transport, such as public bus, at that moment. The practice takes the name of ticket crossing and had spread from Nord Europe to Italy as early as 2011.

In respect of the other emergencies listed above, social behaviors and interactions have no direct impact on physical structures or objects. Anyway from them different and multiple systems of social relation and collective actions emerge.

5 Conclusions: Knitting the Identity Traces

The three examples above discussed show how the identity of the city evolves on the basis of the emergent properties of the system. According to Minati [6], we consider "identity" as an

> emerged property continuously acquired, rather than possessed [6]

by the city. The refuse of the identity as a permanent and regular system is also proposed by Bauman [1], which defines it as an unstoppable experimentation and whose attributes are not predetermined.

The new properties of the identity of a city represent elements that forge the essence of the city that better fit what really occurs in the real context and, as a consequence, its identity.

The emergent properties can assume different forms that reflect the same, or similar, characteristics and needs present on the territory. These occurrences are "traces" only apparently separated; on the contrary, since these single elements are part of a more complex structure, such as identity, they can be reciprocally linked.

Therefore, it is fundamental to identify tools and methodologies enabling, not only the identification and the monitoring of such events, but also their integration (knitting).

In such a way, the identity is built as a structure resulting from elements characterized by continuity in time and space. These elements are repeated on a regular basis in the system, even if they originate from distant fields.

So, the emergencies of the system become acquired properties that transform and develop the city keeping the coherence because of the emergent identity that accompany these processes.

References

1. Bauman, Z. (2003). *Intervista sull'identità*. Roma-Bari: Editori Laterza.
2. Bertrand, D. (1985). *L'espace et le sens*. Paris/Amsterdam: Hadès-Benjamin.
3. Cohen, A. P. (1986). *Symbolising boundaries: Identity and diversity in British cultures*. Manchester: Manchester University Press.
4. Donnan, H., & Wilson, T. M. (1999). *Borders: Frontiers of identity, nation and state*. Berg: Oxford England.
5. Human Smart Cities. (2013). *The Human Smart Cities Manifesto*. Accessed 28 July 2014. http://www.peripheria.eu/blog/human-smart-cities-manifesto
6. Minati, G. (2008). Systemics and emergence for Architecture. In memory of Professor G. Ciribini. Gianfranco Minati, Italian Systems Society. www.AIRS.it. Doctoral lecturer on systems science. Polytechnic University of Milan, Department "Building Environment Sciences and Technology". Accessed 28 July 2014. http://www.slidefinder.net/2/2010_slides/2010-slides/20958787
7. Pezzini, I. (2004). Un approccio semiotico allo studio dello spazio nella città. In F. Martinelli (Ed.), *Città e Scienze umane* (pp. 257–264). Napoli: Liguori.

Chapter 18
Architecture and Systemics: In Praise of Roughness

Carlotta Fontana

1 Introduction

In the first Book of his monumental, ultimate theoretic work[1] about the principles of architectural and urban design, Christopher Alexander identifies fifteen structural features which

> appear again and again in things which do have life.

These properties are: levels of scale, strong centers, boundaries, alternating repetition, positive space, good shape, local symmetries, deep interlock and ambiguity, contrast, gradients, roughness, echoes, the void, simplicity and inner calm, not-separateness [1]. The property of ROUGHNESS seems to me a promising starting point on the way of clarifying the possible links between Architecture and Systemics.

2 Roughness

Roughness, according to Alexander, is an essential structural characteristic of things which have real life because these things "have a certain ease" which prevents them from being morphologically perfect and thoroughly regular. Roughness is not a residue of technical flaws or manufacturing inaccuracy: it is

> an essential structural feature without which a thing cannot be whole.[2]

[1] C. Alexander [1].
[2] C. Alexander, Op. Cit. Book 1, p. 210.

C. Fontana (✉)
Dipartimento di Architettura e Studi Urbani (DAStU), Politecnico di Milano, 20133 Milano, Italy
e-mail: carlotta.fontana@polimi.it

G. Minati et al. (eds.), *Towards a Post-Bertalanffy Systemics*, Contemporary Systems Thinking, DOI 10.1007/978-3-319-24391-7_18

Provided that the industrial idea of optimization is eliminating uncertainty and flaws from the product, Alexander claims that in architecture "life" and "wholeness" have nothing to do with flawlessness. Rather, the act itself of making things gives them life, as

> process is the key to making life in things.[3]

More than that, roughness becomes an inherent, essential quality of architecture, and it should be pursued by design at all scales.

In public spaces, Alexander explains, "the power of instincts" encourages people to take up positions from which they "can protect their backs". He says that, in a courtyard as well as in a public square, "something in the middle"—such as a tree, a monument, a seat, a fountain—is a necessary feature precisely because it allows people to feel more protected.[4] Alexander argues that these centrepieces should be placed according to the pattern of the natural lines of communication which cross a public space, as traced by the people's movement, as in the teaching of the great Viennese planner Camillo Sitte.[5] On this point, Alexander makes a significant observation, noting Sitte's critique that

> the impulse to centre something *perfectly* in a square is an "affliction" of modern times.[6]

That is the reason why the title of this chapter of Alexander's great treatise is: "Something *roughly* in the middle".

At the other end of the scale we find the design of a component, such as a window. Windows are capital transition points in a house, connecting indoor and outdoor space, letting air and light in, keeping cold weather and rain out. In windows, different parts open and close, and different materials meet and connect with each other, according to functional as well as to aesthetic requirements. Industrial components, and the modern system building, normally guarantee the functional quality and the "perfect fit" required. Alexander argues that

> the precision of the component can only be obtained by the most tyrannical control over the plan

in order to reduce tolerances and inaccuracies, while a natural building should be able to keep adapting to the site all along the construction process. Thus,

> a free and natural building cannot be conceived without the possibility of finishing it with trim, to cover up the minor variations which have arisen in the plan, and during the construction.

In such a way, tolerance can be larger and mistakes on the order of half an inch or more can be allowed. While concealing inaccuracies from variations, trims give life to the building, they make its image richer and whole.

---

[3] C. Alexander, Op. Cit. Book 2, p. 4.

[4] C. Alexander et al. [2], pp. 606–607.

[5] C. Alexander et al. [2].

[6] C. Alexander et al. [2].

> Indeed, within this attitude to building, the trim is not a trivial decoration added as a finishing touch, but an essential phase of the construction (...) [it] is in fact a vital part of the process of making buildings natural.[7]

So, ornamentation goes along with function, within a design approach which is actually a philosophy:

> Totalitarian, machine buildings do not require trim because they are precise enough to do without. But they buy their precision at a dreadful price: by killing the possibility of freedom in the building plan.[8]

Arguably, perfect symmetry does not belong to Nature: the interplay between the well-defined order of natural objects and living things, and the constraints of the three-dimensional space in which they grow and develop, seems to result in roughness as a natural quality. In itself, Nature constantly puts constraints in any developing process, which result in variations of each individual—being it a sea-wave, a crystal, a flower ...—within the boundaries of the customary features of its own kind. Thus, the quality of roughness is not caused by inaccuracy but it is the consequence of a well-defined and necessary order.[9]

3 Conclusion

These considerations refer to the introduction of ideas and words coming from the realm of systemics into the world of architectural design, in the early '60s of twentieth century. In those days, design method was investigated and developed as a discipline in its own right.[10] Morris Asimow, one of the most influential author, in his seminal book *Introduction to Design*[11] describes design as an information process[12] consisting in

> the gathering, handling and creative organizing of information relevant to the problem situation; it prescribes the derivation of decisions which are optimized, communicated and tested or otherwise evaluated; it has an iterative character (...).

The idea of performance stemming from this approach was part of a complex design device aimed at reducing uncertainty in the design/construction process and promoting regularity in the product. Architecture is about life, and—as life itself—is admits and endless variability. A wider idea of performance should take into account that

> the system chooses among equivalent configurations according to opportunities which are not prescribed. Equivalent configurations are such because all of them have freedom degrees

[7] C. Alexander [2], pp. 1113–1114.

[8] C. Alexander [2].

[9] C. Alexander, Op. Cit. Book 1.

[10] G. Broadbent [4].

[11] M. Asimow [3].

[12] G. Broadbent [4] , p. 254

and thus all of them are *allowed to happen*. In this way, the system takes on a unique behavior among infinite possibilities. This is the richness of indetermination, as investigated by emergence processes. (...) Concepts such as correct, precise, comprehensive, rigorous, true-false, exact (...) are inadequate for the systemic complexity.[13]

References

1. Alexander, C. (1980–2002). *The nature of order – An essay on the art of building and the nature of the universe. Book 1 – The phenomenon of life; Book 2 – The process of creating life; Book 3 – A vision of a living world; Book 4 – The luminous ground*. Berkeley, CA: The Center for Environmental Structure.
2. Alexander, C., et al. (1977). *A pattern language*. Berkeley, CA: The Center for Environmental Structure.
3. Asimow, M. (1962). *Introduction to design*. Englewood Cliffs: Prentice-Hall.
4. Broadbent, G. (1973). *Design in architecture*. London: Wiley.
5. Minati, G. (2013). Note di sintesi: novità, contributi, prospettive di ricerca dell'approccio sistemico. In L. Urbani Ulivi (Ed.), *Strutture di Mondo. Il Pensiero Sistemico come Specchio di una Realtà Complessa* (Vol. II, pp. 315–336). Bologna: Il Mulino.

[13] G. Minati [5].

Chapter 19
Emergences in Social Systems: Perceptual Factors, Affordances and Performances in Architecture

Alessandra Cucurnia and Giorgio Giallocosta

1 Perception and Architecture

A fundamental condition of each perceptual phenomena, according to Merleau-Ponty, is the *mediating role of the body*, where the latter means

> a synergistic system – all the functions of which are gathered and connected in the overall movement of being in the world – insofar as it is the coagulated image of existence [16].

It also assumes, as is known, the synesthesia of perceptual phenomena, albeit with relationships of *non-equivalence* between the different senses [16], and between the many sensory *effects*.

In architecture (and more generally in the context of territorial and landscape disciplines) perception does not seem to be distinguishable from the more general *fruitive* factors of organisms, contexts, structures, etc.: the latter being factors in which uses, functional connotations, memories, customs, symbolic and cultural values, etc., converge. In Benjamin [1], the description of the relationships existing between the perception of architecture, its *uses* and *customs*, where the latter (through which the *tactical enjoyment*, or the *use* of the architectural artefacts, is chiefly expressed) *largely determine even the optical reception* (or the perceptual events effectively linked to the structures that shape the built environment), is symptomatic. It is observed that: the perception of architecture

A. Cucurnia (✉)
Dipartimento di Architettura, Università di Firenze, Via della Mattonaia 14, 50121 Florence, Italy
e-mail: alessandra.cucurnia@unifi.it

G. Giallocosta
Dipartimento di Scienze per l'Architettura, Università di Genova, Stradone S. Agostino 37, 16123 Genoa, Italy
e-mail: giallocosta@dada.it

G. Minati et al. (eds.), *Towards a Post-Bertalanffy Systemics*, Contemporary Systems Thinking, DOI 10.1007/978-3-319-24391-7_19

> (...) does not appear to be distinguishable from components linked to the *uses* of the structures, thus placing it moreover in a particular position between those same assumptions that – above all historically – have outlined the characteristics and modes of use of artistic works [9].

Again in Benjamin [1], a clear validation may be observed in this respect; distraction and

> concentration are contrasted in a way that allows this formulation: a man who concentrates before a work of art is absorbed by it; he enters into this work of art the way legend tells of the Chinese painter when he viewed his finished work. In contrast, the distracted mass absorbs the work of art. This is most obvious with regard to buildings [1].

In architecture, therefore, through

> the practices of the use of space, behaviors put into effect, and attributions of meaning, the inhabitants create the life that shapes the designed form – and here too it is worth just mentioning that only through this process does the latter fulfill its *raison d'être* (...). It is like saying that a designed form becomes architecture and therefore achieves its entelechy – in the sense of the condition of perfection of the being that achieves its full potential – only when it is *inhabited by a life* [2].

It results in each tautological assonance between architecture and the social systems it is aimed at (and therefore the methods—and developments—in behavior, fruition and perception of the latter).[1]

In such a connotation, emphasis on the concept of architecture is justified by design approaches, which are in fact widely opposed. Of the latter we can mention, for example, those cases that fall under a sort of professional pathology, Ahp—Architects Hate People syndrome [2], according to which,

> for the afflicted designer (...) the subjectivity of those who inhabit architecture is a disturbing, unpredictable and unreliable element; the inhabitant himself is the bearer of chaos, where instead the order of an object of creation complete in itself should reign [2].

2 Emergences and Perceptual Factors

The foregoing reasoning infers the need for architectural design to anticipate the behaviors of structures, anthropized environments, etc., in keeping with the requirements of the users, and the *emergences* (here understood strictly in keeping with the current connotations of the systemic approach) that may arise in relation to the complex interactions between *elements*, *events*, etc. Of these latter, the perceptual factors, along with others, take on significant roles in the emergences of social systems.

[1] See also: [7, 12].

It is affirmed how, for example, in individuals, every perceptual experience determines a *mnemonic exploration* aimed at searching for similarities with a past history, or in their absence, produces the *creation of a new attractor*, assuming for the latter a meaning in keeping with the language of non-linear dynamic systems: in this sense, the attractors can be thought of as

> points, or a set of points, towards which the trajectories that describe the evolution of the system (brain - *author's note*) converge (...). We can therefore conclude that (in each perceptual event, as far as we are concerned here – *author's note*) the brain "lives" (evolves) covering trajectories in the landscape of the attractors (of which each is a memory of past experiences – *author's note*) [18].

Thus,

> the brain calls into question its entire experiential framework and this gives rise to the meaning of the new perceptual experience, which does not belong (...) to the stimulus (...) but to the context of the redesigned landscape of the attractors, in its ever new entirety (...) The reorganization of the landscape of the attractors, as much as the starting point is the framework found at the time of the perceptual act, introduces elements that are not linked to that framework by a relationship of need, and therefore cannot be deduced or predicted by it, link by link in a logical or casual sequence [18].

Merleau-Ponty characterizes perception as that *terrain*

> in which real meanings originate, a terrain that came before scientific knowledge and philosophical reflection [17].

Even the *mediating role of the body* however, *an inescapable condition for every perceptual phenomenon* [16], makes it an equally plausible opportunity for investigations and speculations: above all assuming the *profound* nexuses of the *mind-body* relationship and the *non-separability* between the *observer* and the *observed object* (or *topic of discussion*, or *theme for reflection*, etc.).[2] The same connotation of *phenomenal body* (or *body-acquaintance*) in fact, strictly ascribable to the existentialist thinking of Merleau-Ponty, gives rise to each legitimization in that sense: the analysis of the perception

> allows access not to a "transcendental field" but a "phenomenal field", to grasp existence in its corporeality and in its immediate relationship with the world experienced, which goes beyond the objective world of science. Merleau-Ponty was far from devaluing the role of scientific knowledge (...) but believed that it was something derived, the result of a detachment from the "world of life" (...). The terrain of his investigation is comprised not of consciousness but by the structures of the "world of life" (...). The outcome of his analysis was the outright rejection of a consciousness-based approach, the origin of which he saw in the Cartesian *cogito* (...). Reinterpreted in phenomenological terms, the *cogito* reveals to us not a separate ego but the existence of man in the world, linked by relationships with his own and with other bodies: consciousness does not exist separately from things, but it always relates to the structures of being-in-the-world [17].

[2] This *non-separability* is nevertheless assumed here in the sense that it does not obliterate further aspects and issues, as for example in: [5].

3 Affordances

The discussion of perceptual topics may shift from various approaches, such as that ascribable to the cognitive sciences, or to *perceptual functionalism*, or that derive directly from Gestalt psychology, etc. (of which, for example, there are a few mentions in: [3]). A radical hypothesis in this sense is that developed by Gibson [11], and still considered the effect of a basic criticism of the cognitive conception.

In a nutshell, and as far as chiefly concerns us here, the hypothesis uses the concept of *affordance*, a neologism derived from the verb *to afford* and assumable as *the possibility of taking advantage of certain opportunities, availability*, etc.: so, *affordances* can be understood as the *availability of objects even with regard to their potential uses*, and as *calls to action*. According to Gibson [11], the *objects* of perception consist of values and meanings *external* to the perceiver and influential on the latter *through the opportunities offered* (*affordances*). In this regard, and in clear disagreement with *orthodox* psychology, Gibson [11] explicitly states that, during the observation of objects, the individual perceives *not the qualities* but rather the *opportunities offered.*

An *affordance simultaneously* concerns the perceived environment as much as the perceiver (or the *observer*) and his body. In fact, according to Gibson [11], we all *adapt to the environment insofar as we are shaped by it*. For this reason, *perceiving the world is to perceive ourselves at the same time* [11].

In Gibson [11], moreover, the concept of environment is made to coincide with that concerning the *world of ecological reality*. This latter is characterized by *objects endowed with meanings* and, as such, *revealed inasmuch as made available* through perceptual experiences. In this sense, for example, the edge of a precipice can be perceived as a *harmful affordance*, the flat surface of a lawn as an *invitation* to engage in certain activities, etc. In contrast to that *of the ecological reality*, the *world of the physical reality* is instead characterized by meanings *imposed* on *objects*, rather than *discovered* [11]. If Gibson's hypothesis were valid, it is observed, the effects in terms of the architectural design culture would be numerous and significant, as it would define

> the dominant mode through which an individual relates to the surrounding space [2].

Nevertheless from other hypotheses, and from the relative convergence factors, significant implications for architectural design and its possible evolutionary developments can be inferred.

4 Perception, Performances and Affordances

Of these implications, the requirements-performance based approach to the design of architecture certainly represents an area of disciplinary interest in speculations concerning perceptual factors (see for example, of the most recent contributions in this regard: Cucurnia [3]). A distinct aspect of this approach is the immanence of a relationship, *tautological* for architecture (despite the more prohibitive deviations, such as the so-called *Ahp syndrome*, already mentioned), with the social systems and their requirement-based dynamics, sometimes explicitly clear and at other times only implicitly underlying and therefore in need of unveiling and interpretations [7, 8].

The relationships existing between perceptual factors and developments in requirements-performance based approaches can be briefly inferred, moreover, even by only considering what is argued, in this regard, about the effects of the former on the latter. As regards the causes that *motivate* a requirement,

> it is necessary to consider the numerous perceptual states that an individual (...) assimilates and records in sequence (...) The occurrences (past and present), that coexist in their interdependence [13], acquire value in relation to the magnitude of the emotions that produce and seem to determine the requirements-based developments. The requirement formation process (...) is therefore connected to the perception of the surrounding events, by learning from the past and with the ability of the individual to *foresee the future* (or identify the most likely scenarios – *author's note*) on the basis of their own experiences framework [3].

The *objective* and *phenomenal* dimensions of the body also converge in the perceptual processes (see, above all in relation to the *phenomenal dimension* of the body: Merleau-Ponty [16]). Each perceptual event, it is affirmed, involves the physical-sensory structure of the perceivers

> but also their psychological and socio-cultural spheres [15]; in fact, in relation to the socio-cultural sphere, we know of the role assumed here by "social processes that culturally mediate the perceptual and learning dynamics" ... [10, 15].

Even following Gibson's hypothesis, the connection between perceptual events and the development of requirements systems is equally confirmed. In this case, in fact, the *affordances*, which act as *objects* of perception as we have already observed, also perform roles of unveiling—or implementing, or *validating*—certain needs. In this regard it is sufficient to consider how *recognition* of the *opportunities perceived* cannot help but pertain to the *usefulness* found in them, and therefore the latter's obvious connection with the aim of satisfying *those* needs that motivate it: the connotation of *usefulness* naturally remains even in the case, for example, of *harmful affordances* (from which the danger levels of certain environments, objects, etc. are generally inferred).

Here however, and particularly obvious in the case of architecture, the modes with which *an individual relates to the surrounding space* [2] defined through the perception of the *affordances*, would express the synchronicity between the requirements that have *emerged*—if previously non-explicit—and their satisfaction (or in any case the identification of the strategies for the pursuit of the latter). Gibson's hypothesis, applied to architecture, may also be understood as enhancing the relationships between designers and users: here however, risks may arise of misunderstandings about the evolutionary developments of the requirements-performance based approach to the design [6], or it may reveal reductive connotations.

Where for example the latter—and especially its developments—means not appreciably permissive of poietic attitudes on the part of users concerning the organization of their living places, the design of the *affordances* of anthropized environments, rather than their *performances*, would seem more in keeping with allowing the possibility of essentially optimal uses in terms of the personalization of spaces, buildings, etc., by the perceivers (here strictly understood in the sense explained by Gibson).[3] In reality, it is thanks to the perception of the designed space (here too strictly maintaining Gibson's hypothesis), and above all the *quality* of this space conferred to it by the design concerning:

- the expectations of the users (explicit and implicit, current and future),
- the relative *affordances* consistent with them,

that results in *scenarios* where poietic attitudes and the personalization of living areas by users are made possible (or at least *facilitated*) as they are *part of those* opportunities offered, *and not others*. Therefore recognizing that the prerogative of the developmental lines of the requirements-performance based approach is to consider the *many* types of *needs and requirements to be met* (functional, cultural, poietic, etc.), the *performances offered* by an architectural project, so decisively oriented, do not clash with consistent connotations of *affordances*, but rather exhibit significant conceptual assonances.

In this sense, even where there is emphasis (as already mentioned) on the *practical uses of the space* which would represent the *method* through which *the inhabitants create the life that shapes the designed form* [2], for the latter (and more generally for the *performances* equated with the *designed form*) it may reflect a greater propensity to *being inhabited by a life* in keeping with the *expectations that dynamically arise within it*, as far as the project is geared towards those expectations. Similarly, if through

> the expression of their intentions of appropriation and use, the inhabitants turn what is only a *space into an opportunity* (...), a *space implemented as reality* [2],

the use of the latter becomes all the more optimal the more the design of the former influences the perceivers with *affordances* geared towards satisfying their needs.

5 Conclusions

The *emergences* in social systems, here assumed in their connection with perceptual processes, clearly concern other areas (behavioral, phenomenological, etc.).[4] Likewise for the former there is a need for further development and interdisciplinary contributions—strictly attributable to a systemic approach—above all aimed at:

[3] Generally speaking, in architecture, *performances* can be taken as the *behaviors* (under certain conditions of use and strain) of spaces, elements, and structures, in relation to the requirements to be met. See also, on the concepts of requirement, requisite and performance: Maggi [14]; succinctly: Cucurnia [3].

[4] See for example: [4, 7].

- advances in knowledge on the relationships between perception and needs,
- the definition of operational guidelines in keeping with the knowledge acquired.

What has been hitherto discussed is intended as a contribution in this sense.

References

1. Benjamin, W. (2000). *L'opera d'arte nell'epoca della sua riproducibilità tecnica*. Turin: Einaudi (original edition 1936).
2. Chiesi, L. (2009). Lo sguardo dell'architetto. Gli indizi ambientali fra scienze sociali e progettazione. In G. Amendola (Ed.), *Il progettista riflessivo*. Bari: Laterza.
3. Cucurnia, A. (2014). Percezione ed evoluzioni di approccio esigenziale-prestazionale. In G. Giallocosta & A. Magliocco (Eds.), *Fattori percettivi in architettura*. Florence: Alinea.
4. Di Battista, V., Giallocosta, G., & Minati, G. (2010). L'auto-architettura nei sistemi sociali. In P. Capone (Ed.), *Ricerche ISTeA verso una edilizia "ragionevole"*. Florence: Medicea.
5. Giallocosta, G. (2009). Systemic and architecture: Current theoretical issues. In G. Minati, M. Abram, & E. Pessa (Eds.), *Processes of emergence of systems and systemic properties* (pp. 91–100). Singapore: World Scientific .
6. Giallocosta, G. (2011). Tecnologia dell'Architettura e progettazione tecnologica, *Techne, 2*, 24–31.
7. Giallocosta, G. (2012). Architecture and development of and within social systems. In G. Minati, M. Abram, & E. Pessa (Eds.), *Methods, models, simulations and approaches towards a general theory of change* (pp. 495–503). Singapore: World Scientific.
8. Giallocosta, G. (2013). Progetto e percezione: criticità e prospettive. In G. Scudo (Ed.), *Tecnologie solari integrate nell'architettura*. Milan: Wolters Kluwer Italia.
9. Giallocosta, G. (2014). Architettura, fruizione, percezione. In G. Giallocosta & A. Magliocco (Eds.), *Fattori percettivi in architettura*. Florence: Alinea.
10. Giallocosta, G., & Piccardo, C. (2014). Integrazione architettonica di tecnologie da fonti rinnovabili: rapporti con i fattori percettivi e orientamenti per linee-guida operative. *Techne, 7*, 165–170.
11. Gibson, J. J. (1979). *The ecological approach to visual perception*. Boston: Houghton Mifflin.
12. Gregotti, V. (2008). *Contro la fine dell'architettura*. Turin: Einaudi.
13. Lewin, K. Z. (1951). *Field theory in social science*. New York: Harper & Row.
14. Maggi, P. N. (2009). *Il processo edilizio*. Milan: Polipress.
15. Mela, A., Belloni, C. M., & Davico, L. (2002). *Sociologia dell'ambiente*. Rome: Carocci.
16. Merleau-Ponty, M. (1945). *Phénoménologie de la perception*. Paris: Gallimard.
17. Rossi, P., & Viano, C. A. (1999). *Storia della filosofia – Il Novecento*. Bari: Laterza.
18. Vitiello, G. (2010). Dissipazione e coerenza nella dinamica cerebrale. In L. Urbani Ulivi (Ed.), *Strutture di Mondo. Il Pensiero Sistemico come Specchio di una Realtà Complessa* (Vol. I, pp. 105–126). Bologna: Il Mulino.

Chapter 20
Bank of Experiences: A Tool to Enhance Creativity, Enterprises and Countries

Giordano Bruno and Giulia Romiti

1 Introduction

Bank of Experiences (BdEsp) is an job agency that puts creative potentials of engineers, designers, experts of marketing and business organization (and, more generally, professionals related to the "world of research in science and design") in touch with companies looking for active collaborations for process or product innovation. BdEsp is a project of design management based on maps of system and of infographics, which identify the visual aspect of the platform, that is an interactive system of storage and organization of the Bank's data and exchange flows.

The aims of the BdEsp are the management and development of human resources, real value of the Bank (Figs. 20.1 and 20.2).

2 Chapter 1

The aims of the BdEsp are the management and development of human resources, real value of the Bank.

Within the development of a post Von Bertanlaffy systemic, BdEsp represents a contribution of exemplifying innovative themes such as the ability to realize multiple and dynamical coherence, the use of formal ontologies and maps with representations through fractal structures, the ability to induce new properties, the use of structural dynamics (Fig. 20.3).

Users access the virtual BdEsp platform, adhering to the goals of social economy that are foundation to the system. These conditions make the project based

G. Bruno (✉) · G. Romiti
ISIA Roma Design, Istituto Superiore per le Industrie Artistiche, Roma, Italy
e-mail: gibrun84@gmail.com; giulia.romiti@isiaroma.it

G. Minati et al. (eds.), *Towards a Post-Bertalanffy Systemics*, Contemporary Systems Thinking, DOI 10.1007/978-3-319-24391-7_20

Fig. 20.1: Introductive map of systemic open

on well-defined values, placing a strong emphasis on the fundamental principles of system design. Great relevance for the system is, above all, the principle of gratuitousness, understood as a way of life and interpersonal relationships, recently well described in the research of the Italian economist Luigino Bruni [1]. Another key concept is the degrowth, theorized by philosopher and economist Serge Latouche; according to the theory of degrowth, real wealth does not consist in material goods and consumption but in human relations that create surplus value also for the economy [5].

One of the main point of the system is the acceptance of the users belonging to the platform; this stage, set in its main flows by the system managers, has the goal to bring out skills and relational/job qualities of the users.

User will be entirely and actively involved in the system, from the drawing of a indicative format to the creation of a dynamic, continuously evolving chart, thus becoming an actual part of its organization and management.

The graph shows, at the same time, skills already acquired and those to be integrated in a continuous process of feedback. In the second case it is suggested a training program, settled up by those companies and professionals, already belonging to the Bank, chosen for the role of tutors. The system, indeed, has lifelong learning, spread of systemic culture and values such as sharing and collective welfare as its primary objectives . The basis of both system profile's analysis and creation of work teams is the principle of "Five kinds of minds for the future", theorized by psychologist Howard Gardner [2]. According to Gardner, indeed, it is necessary to develop five intelligences that will be needed for the future of the

Fig. 20.2: Inside the introductive map

company; this differentiation is critical to the BdEsp system, which aims to create heterogeneous research teams, thus changing the service offered to companies and the training provided to its users (Fig. 20.4).

Within Bank work-flows, the creation of teams according to different relationships (specific to those researches which are supported and determined by BdEsp) is strongly encouraged. These flows are able to generate multiple coherence and dynamics that determine the consistency of the system. Through interactions with the system, users establish local interactions to each other, which are able to create a global, wider and more widespread behavior .

3 Chapter 2

Managers control the team's constitution, supporting the users in the aggregation to profiles which can set up very versatile work-teams. This mode of operation is implemented by companies and research centers in order to conduct their experimental projects.

Companies adhering to the Bank will be able to initiate research projects or to make teams work on their existing/to be implemented projects. Activities will follow specific time-lines and process designs, which will provide opportunities to meet and exchange—not only in "virtual-mode". BdEsp will support the participating

Fig. 20.3: PDF Map of the actors of the system

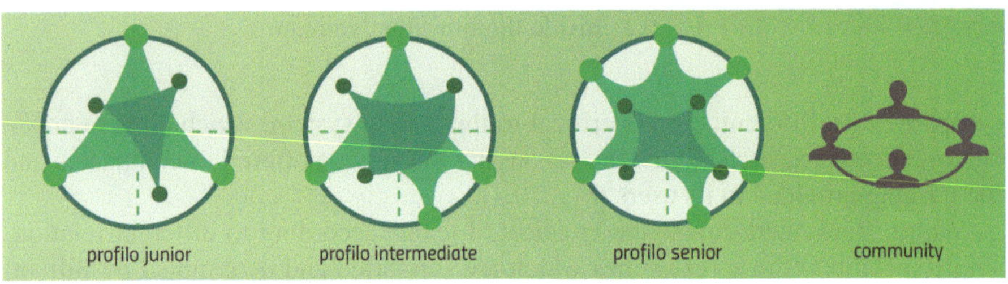

Fig. 20.4: Graph of ability

companies during the research and the management team will play an important role in mediating between the actors involved in these projects (Fig. 20.5).

Following these operating modes, the companies can take advantage of permanent consultancies by experts, get in touch with other companies belonging to the platform and implement joint-venture projects. The users of the Bank will be able to conduct research and acquire work experience at the same time, with a chance of enhancing their own skills. As for the platform's managers, the system is a source of managerial and systemic experience, providing them job opportunities and professional growth.

The "support team" play the role of a facilitator and are made by the users of the system who are called into the management of the platform. This team plays a key role: they control the work of the teams of the Bank by giving skill points and help enhance competence of users' profile. They also have the function of keeping the

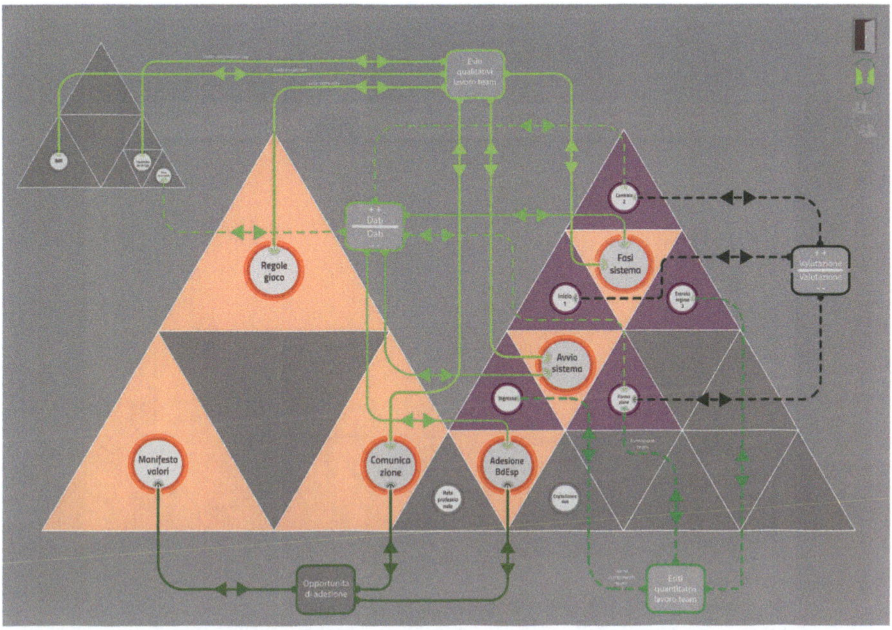

Fig. 20.5: Maps of workflows

system alive and open to exchanges with external-reality. Another important goal of the team is promoting the activities of the operators, providing critical evaluative feedback on the service provided.

Support team and managers take advantage of those system nodes that—as qualitative and quantitative filters measuring profile's skills and the number of participants adhering—provide feedback in order to adjust both the system and the work of its actors (homeostasis of the system).

4 Chapter 3

BdEsp project, then, as an online and "on-earth" platform, represents a fertile ground for collaboration between companies and communities for the development and support of wide networks of active exchanges. The Bank creates interactions and cooperation in a view of an open system addressed both outside (open to local networks and/or different companies) and inside (territorial district and/or individual business). The values relevant to the concept of sharing and of social economy encourage relationships between the actors and the good practices of dialogue and exchange (Fig. 20.6).

As for the informatics part, BdEsp contemplates the use of ontology, which is a method of classifying information according to specific formal representations

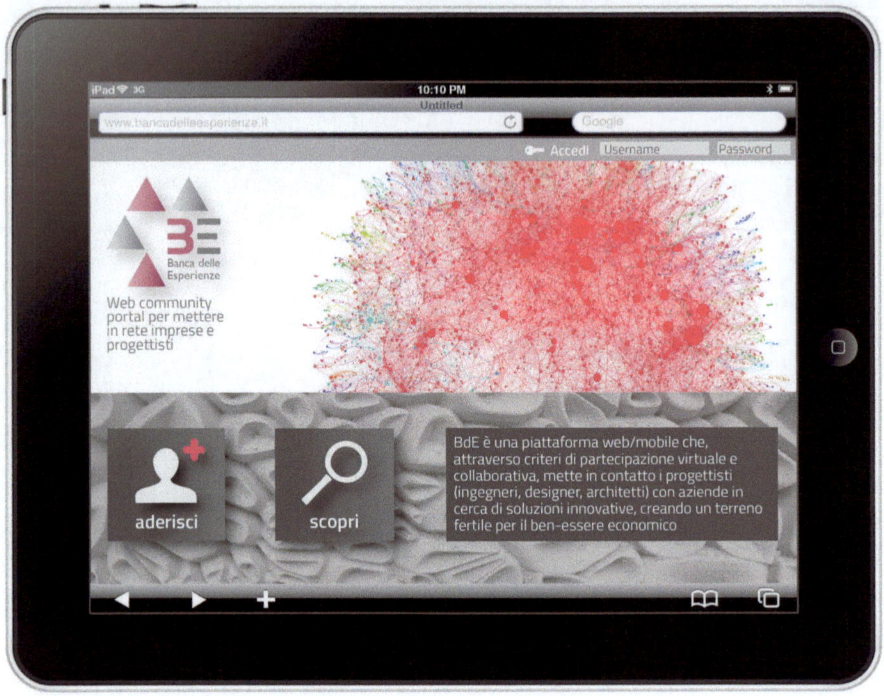

Fig. 20.6: Simulation of application interface

of the data (formal ontology). With this type of formal organization, a hierarchy of information is underlined, in which the relationships between the actors of the system are described .

Thanks to the concept of ontology is also possible to deepen the concept of metadata, which is vital in the design of the Bank of experience. Metadata, i.e. data that describes data, are used to classify information with respect to certain classes or concepts. This is exactly the process that occurs during the observation and evaluation of users' profiles in entrance, which are properly organized to enable a better development of teamwork and operation of the platform.

5 Conclusions

The system was represented through conceptual maps, able to communicate its complexity. From the graphical point of view—and in line with the criteria of development and system's growth—we have chosen a fractal pattern (Sierpinski triangle, also called perpetual triangle) declined in several aggregating and graphical ways.

Fractal geometry, which characterizes the system in the graphical and conceptual shape, also fits to scalability logic. According to these logic, the system is not limited to a single level but can be observed in more degrees of depth (from the reception phase to the operational teamwork's process) and using different approaches. The main objective of the scalability logic is to ensure proper management of system resources at all levels of the project. The triangular matrix pattern allows to communicate those aspects of connection and exchange among the actors of the project according to sizes and color gradients, useful to clearly define the roles of the actors involved in the early stages of the system.

Acknowledgements Thanks to Veneranda Carrino and Mario Fois who support the project.

References

1. Bruni, L. (2009). *L'impresa civile. Una via italiana all'economia di mercato*. Milano: Università Bocconi.
2. Gardner, H. (2007). *Cinque chiavi per il futuro*. Milano: Feltrinelli Editore.
3. Guarino, N. (2013). Services and Service Systems under a Mesoscopic Perspective. In E. Gummesson, C. Mele, & F. Polese (Eds.), *Service dominant logic, network and systems theory, and service science: Integrating three perspectives for a new service agenda*. Napoli: Giannini.
4. Guarino, N., & Welty, C. (2000). A formal ontology of properties. In *Knowledge Engineering and Knowledge Management Methods, Models and Tools, 12th International Conference, EKAW 2000, Proceedings, Juan-les-Pins, France, 2–6 October 2000. Lecture notes in computer science* (pp. 97–112).
5. Latouche, S. (2005). *Come sopravvivere allo sviluppo. Dalla decolonizzazione dell'immaginario economico alla costruzione di una società alternativa*. Torino: Bollati Boringhieri.
6. Minati, G. (2008). *New approaches for modelling emergence of collective phenomena. The meta-structures project*. Milan: Polimetrica Editore.
7. Minati, G. (2011). Quale conoscenza per gestire la complessità della società della conoscenza e dei suoi processi? *Riflessioni Sistemiche, 5*, 122–137.
8. Minati, G. (2012). Scelte manageriali: dal far funzionare dispositivi sociali all'emergenza, Reti, Saperi, Linguaggi. In I. Licata (Ed.), *Sistemi, Modelli, Organizzazioni. Management e Complessità* (pp. 15–38). Roma: Corisco.
9. Minati, G. (2013). Note di sintesi: novità, contributi, prospettive di ricerca dell'approccio sistemico. In L. Ulivi (Ed.), *Strutture di mondo. Il pensiero sistemico come specchio di una realtà complessa* (Vol. 2, pp. 315–336). Bologna: Il Mulino.
10. Minati, G., & Pessa, E. (Eds.). (2002). *Emergence in complex, cognitive, social and biological systems*. New York: Kluwer Academic/Plenum Publishers.

Part VI
Outlines of a New General Theory of Systems

Chapter 21
Systems and Organizations: Theoretical Tools, Conceptual Distinctions and Epistemological Implications

Leonardo Bich

1 Introduction

What is a system, and how can we characterize and identify it with respect to its background? Which are the relations and components that are crucial for its description? Every domain of investigation makes different distinctions when identifying a system, and it is not just a question of scale: different kinds of relations are considered as pertinent in order to describe the phenomena or object of study, and different operations of partition are performed in order to extract the relevant components. Let us think for example of how many system domains can be found in a human body: from molecular and cellular ones, to systems including complexes of organs up to ecosystems populated by our bacterial symbionts. As a consequence, the same material entity can in principle be described in terms of different kinds of systems, each with specific components and organization.

Indeed, the word "system" is almost never used alone, but it is usually paired with an adjective that specifies its domain of application: physical, chemical, biological and so on. One of the challenges is to develop as much as possible the understanding of the relational dimension that characterizes "systemhood" independently of realizations in specific domains [20, 241].

In this respect, of course, the main focus is on the notion of organization. Defined generally as the topology of relations that characterizes a certain system, it can mean very different things, and its specification could be somehow arbitrary or, however, extrinsic to the system, to the extent that it would depend on the purposes of an external observer or designer. Although such specification might be useful in the

L. Bich (✉)
Department of Logic and Philosophy of Science, IAS-Research Centre for Life, Mind and Society, University of the Basque Country (UPV/EHU), Avenida de Tolosa 70, 20018 Donostia-San Sebastiàn, Spain
e-mail: leonardo.bich@ehu.es

G. Minati et al. (eds.), *Towards a Post-Bertalanffy Systemics*, Contemporary Systems Thinking, DOI 10.1007/978-3-319-24391-7_21

domain of artifacts, it might be of little help in the case of natural and human systems. What I suggest is to find ways to characterize organization endogenously, in terms of the degree of functional integration[1] achieved by a system, that is in terms of an effective role played by a specific topology of relations in specifying the internal and external dynamics of the system. This can allow us to identify a system according to what is relevant to its own dynamics, and not just as the result of an arbitrary, or theoretically weak partition of a certain *medium*.

One way of doing this is to focus on those systems—such as living ones, but not only them, let us think also of ecological and social ones—that are capable of forms of self-maintenance, that is, of specifying to a certain degree their own dynamics and maintain themselves with respect to their environments. In this context a thriving work has been recently done based on the concept of constraint, which implications I will present and analyze in the following sections.

2 The Notion of Constraint and Its Organizational Implications

A crucial role in characterizing natural systems, especially self-maintaining ones, was ascribed to constraints by pioneering researchers in System Theory, such as Howard Pattee [17]. Yet, it is especially in the last two decades that this notion has raised a renovated interest and has undergone a profound development [8, 9, 13].[2]

Usually the term constraint stands for an asymmetrical relationship such as that holding between boundary conditions and dynamics. When the behavior of the system is underspecified, constraints constitute an alternative description which provides the missing specifications (normally by decreasing degrees of freedom). The pivotal role played by this notion in Systems Theory depends on the fact that it allows us to focus not only on the internal dynamics of a system, but also to take into consideration the conditions of existence of these dynamics, and how in some cases they can be affected by the activity of the system itself: general examples are the river modifying its bed, or a living system modifying the boundaries conditions of its internal environment (Ph, osmotic pressure, concentrations of enzymes, etc.).

Speaking of properties of the internal environment, a foundational role in this tradition had been played by Claude Bernard's pioneering work already in the middle of the nineteenth century. He distinguished between natural laws, common to all phenomena, and *mileux*, those boundary conditions that specify the specific properties of distinct phenomena [1]: different *mileux* realize distinct phenomena, not because they follow different natural laws, but because they are characterized by different sets of constraints acting in addition to laws.

[1] By functional integration I mean here the degree of mutual dependence between those subsystems and processes—what I would call functional as opposed to structural components [2]—that are necessary for the functioning of a system and are identified and characterized in terms of such contribution.

[2] See Umerez and Mossio [22] for a brief but detailed review on the notion of constraint.

Bernard applied this very powerful tool to the case of organisms. Living processes exhibit distinctive properties with respect to other natural systems due to the specificity of their internal *milieu*. Their internal *milieu*, in fact, is self-produced, self-specified, and self-maintained, since all components contribute to the realization of the conditions in which all other components are immersed.

The underlying idea is that in some way living systems are capable of generating as well as maintaining some of their distinctive constraints. And this idea is at the basis of a notion of organisms as full-fledged systems: unities distinguishable from their environment in terms of their own activity, and whose organization plays an effective role in specifying their underlying dynamics.

Yet, the notion of constraints has always escaped precise definition, besides the general acknowledgement of its role in providing additional specifications to dynamics that otherwise would be insufficiently (or incorrectly) described. With the goal of providing a naturalized notion of constraint capable of expressing operationally its role within a system—not only as an independent external condition—a definition has been recently proposed:
Given a particular process *P*, a configuration *C* acts as a constraint if

1. at a time scale characteristic of *P*, *C* is locally unaffected by *P*;
2. at this time scale *C* exerts a causal role on *P*, i.e. there is some observable difference between free *P*, and *P* under the influence of *C* [14, 164].[3]

Typical examples are the activity of an enzyme, which catalyses a reaction without being directly affected by it; a pipe harnessing a flux of water, etc.

The relevance of this definition lies in the fact that (1) it specifies and allows us to describe two orders of "causes" in natural systems: processes and constraints; (2) it entails a notion of organization that is more complex than a flat network of relations, by introducing a basic functional hierarchy; and (3) it allows us to characterize constraints both in terms of their composition and realization (as material structures), and in terms of their action upon lower level material processes (as functional components of a system). In the following two sections I will present some implications of this idea, and I will propose some conceptual distinctions based on it.

3 Two Kinds of Organizational Closure

The notion of constraint has been recently used to describe a fundamental feature of (biological) self-maintaining systems, that is, their circular organization through which the activity and existence of the system come to coincide.[4] This idea has been

[3] A more detailed analysis can be found in [12].

[4] The idea that in those far from equilibrium systems which are capable of self-maintenance and self-production, the very existence and activity of their constituents depend on the network of processes of transformation that they realize, and they collectively promote the conditions of their own existence through their interaction with the environment.

expressed in the literature through the notion of (organizational) *closure* (by Piaget [18], Rosen [19, 21], Maturana and Varela [10]). Yet, by means of the concept of constraint closure can be expressed more rigorously in such a way as to embrace not only self-production and self-maintenance, but also the contribution of the system to its own conditions of existence. The basis of this reformulation of closure derives from Bernard's notion of internal milieu, and it is implicitly alluded in the autopoietic theory when it emphasizes the role of the membrane of a living cell in contributing to the specification of its self-determined internal phase space. The idea consists in taking specifically into account the capability of the organization of a system subject to closure to specify part of the internal and external boundary conditions that enable and control its dynamics. The result is the possibiity to characterize system that is capable of a minimal form of self-determination, rather than being driven by external conditions.

Starting from a conceptual reformulation of Kauffman's [9] idea of work-constraint cycles and Rosen's [21] model of closure to efficient causation, it is possible to characterize organizational closure along this line as a closure of constraints [15]. In this view a system realizing closure is capable of generating some of the constraints that control and enable its dynamics, in such a way that the existence and activity of each of these constraints in turn depends on the action of other constraints in the system. Therefore closure consists in a mutual (generative) dependence between self-produced constraints acting on basic processes.[5] An example is represented by Kauffman's abstract auto-catalytic sets, where all the catalysts (i.e. constraints) are produced within the system through the contribution of other catalysts in the system, acting as constraints on the underlying biochemical processes. By expressing closure in terms of constraints, this approach is able to provide a precise characterization of what is considered as functional closure (at the level of constraints) as opposed to physical closedness (e.g. the consequence of a boundary), or to structural openness (at the level of processes, the flux of environmental matter and energy on which the system acts to maintain itself).

This idea is also very useful in order to distinguish between two different uses of the word closure, *operational* and *organizational*, often confused with each other. As stated in [4], a fundamental difference between them lies in the fact that the former implies a form of recursivity of operations, while the latter has a deeper self-referential character (which can be expressed globally by a self-referential function $f(f) = f$ rather than a recursive one). Organizational closure, unlike operational one, involves not just a circular recursion or a closed network of operations, but rather a mutual generative dependence between components realized through a closed topology of transformation processes.

What is crucial besides the activity of the components is the status of their conditions of existence. This distinctive feature of organizational closure becomes clearer, and distinctions can be made more precisely, if we express it in terms of constraints:

---

[5] For each constraint C_i, (at least some of) the boundary conditions required for its maintenance are determined by the immediate action of another constraint C_j, whose maintenance depend in turn on C_i as an immediate constraint. The system is self-maintaining because its constraints, through closure, are able to act on some dynamics in such a way that, in turn, the same dynamics contribute to maintain some of the boundary conditions that allow their existence [12, 14, 15].

we have operational closure when there is a circularity of processes (e.g. exchanges of signals in a network of computers), but all the constraints that enable it are independent from it and externally specified. There is organizational closure, on the other hand, when some of the constraints are produced from within, i.e. when at least part of the conditions of existence of the organization are specified by the very dynamics of the systems (through mutually dependent functional components acting as constraints).

Therefore, introducing the notion of constraint in the characterization of systems provides a powerful theoretical tool, which makes it possible to make distinctions between hetero-specified and self-specified organizations.

4 Levels of Integration

How can distinct constraints be integrated in a system organization? And in what sense and to which degree can we say that they are mutually dependent? Let us consider here two simple cases of model systems that achieve self-maintenance by realizing closure: the Chemoton [7], and M/R-systems or auto-catalytic sets [9, 19, 21]. Both are characterized by hierarchical networks involving two orders of causes (processes and constraints), but they realize two different forms of systemic integration, that we can define respectively "confederative" and "unitary" [2].

Let us think first of Ganti's Chemoton, a model of pre-biotic system organized as a biochemical clockwork, in which three autocatalytic subsystems—respectively a metabolic cycle, a template subsystem and a compartment—are directly coupled like chemical cogwheels. The autocatalytic subsystems act as constraints on the underlying biochemical fluxes, and interact with each other in terms of supply and demand of metabolic substrates.

These subsystems are mutually dependent—and therefore realize closure—only in a very simple form, to the extent that they provide one another the material substrates necessary for their own maintenance. But in principle they could exist in isolation, provided the environment contains the appropriate nutrients in the right amount.

Let us consider, in comparison, Rosen's and Kauffman's models. Both are metabolic networks characterized by organizational closure in presence of cross-catalysis: that is, each catalyst is generated through the action of at least another catalyst, which constraints the process of production of the former, in such a way that they are collectively capable to realize self-production and self-maintenance. In this case different constraints are not just simply coupled through supply and demand of metabolites, but each depends on the direct action of another constraint for its production and maintenance.

By considering the relation between constraints, therefore, it is possible to identify different forms of functional integration even in very basic systems realizing organizational closure. The Chemoton represents the most basic degree of integration, that we can call integration of level 1, between coupled constraints; in the second case a generative dependence establishes a level of integration 2 between mutually enabling constraints.

A new degree of integration (of level 3), in turn, emerges in presence of mechanisms of coordination of basic functions (such as regulatory ones), that is, in presence of new orders of constraints that independently modulate the underlying ones, by selecting between different basic functional regimes available [5]. The hierarchy can grow further by adding new functional orders.

5 Epistemological Remarks: Constraints and Degrees of Logical Openness

An analysis in terms of constraints (and forms of self-constraint) conveys a strong notion of system, that is, a self-specifying unity with a highly integrated organization. A first epistemic implication of it concerns the status of components. The idea that they depend on the system for the specification of their behavior and, even more, that they also exist only as far as they are part of the organization, implies that they have to be identified (as constraints), with respect to the role they play in this very organization, that is: *top-down* as functional constraints, rather than bottom up on the basis of their material composition [3].

Another and more general implication is related to the fact that the constraints considered in the previous sections are *non-holonomic* [17], that is, they are themselves dynamical, time-dependent and therefore nonintegrable, and they realize a (indirect) loop with the dynamics they affect. On this basis I suggest that a correspondence can be established between orders of constraints and degrees of logical openness [11],[6] as each new order of constraints poses further limitations to the possibility of providing a dynamical description (see for example Hooker [8]).

Basic closure, i.e. one order of self-constraint, would exhibit a logical openness of degree 1, since there are already limitations to the possibility of its dynamical description, and alternative strategies of description are required. For example simulations [6], though providing only partial descriptions, are better suited for catching its distinctive features, and synthetic realizations would be even more informative.

Regulatory mechanisms, as higher-order constraints, add further degrees of logical openness (2 or more), and we know that natural complex systems, unlike basic simplified models such as those analyzed here, are characterized by many more interacting orders of constraints acting at different levels of organization. In such a scenario, each phenomenon would require different modeling strategies as well as specific criteria for selecting which are, functionally speaking, the most pertinent levels of organization involved and the relative constraints: describing ecosystems would imply considering, for example, the set of constraints directly involved in the relation between organisms and niches, rather than those at the level of the cells that compose these organisms [16]. Therefore when multiple orders of constraints are involved, the application of specific selective criteria and the combination of

[6] The possibility or not to formulate models of the behavior of the system that converge to an optimal (or complete) description of it.

qualitatively different, though partial, models—chosen heuristically, according to the goals of the explanation—seems the most fruitful alternative strategy to that, impracticable, of building increasingly comprehensive dynamical models.

Acknowledgements This work was funded by the Basque Government, Spain (Postdoctoral fellowship and research project IT 590-13) and by Ministerio de Economia y Competitividad, Spain (Research project FFI2011-25665).

References

1. Bernard, C. (1865). *Introduction á l'étude de la médecine expérimentale*. Paris: Baillière.
2. Bich, L. (2010). Biological autonomy and systemic integration. *Origins of Life and Evolution of Biospheres, 40*, 480–484.
3. Bich, L. (2012). Complex emergence and the living organization: An epistemological framework for biology. *Synthese, 185*, 215–232.
4. Bich, L., & Arnellos, A. (2012). Autopoiesis, autonomy and organizational biology: Critical remarks on "Life After Ashby". *Cybernetics and Human Knowing, 19*(4), 75–103.
5. Bich, L., Mossio, M., Ruiz-Mirazo, K., & Moreno, A. (2015). Biological regulation: controlling the system from within. *Biology and Philosophy*.
6. Cornish-Bowden, A., Piedrafita, G., Morán, F., Cárdenas, M.L., & Montero, F. (2013). Simulating a model of metabolic closure. *Biological Theory, 8*(4), 383–390.
7. Ganti, T. (2003). *The principles of life*. Oxford: Oxford University Press.
8. Hooker, C. (2013). On the import of constraints in complex dynamical systems. *Foundations of Science, 18*, 757–780.
9. Kauffman, S. (2000). *Investigations*. New York: Oxford University Press.
10. Maturana, H., & Varela, F. J. (1973). *De Máquinas y Seres Vivos: Una teor'a sobre la organización biológica*. Santiago: Editorial Universitaria.
11. Minati, G., Penna, M. P., & Pessa, E. (1998). Thermodynamical and logical openness in general systems. *Systems Research and Behavioral Science, 15*(2), 131–145.
12. Montévil, M., & Mossio, M. (2015). Biological organisation as closure of constraints, *Journal of Theoretical Biology, 372*, 179–191.
13. Moreno, A., & Mossio, M. (2015). *Biological autonomy: A philosophical and theoretical enquiry*. Dordrecht: Springer.
14. Mossio, M., Bich, L., & Moreno, A. (2013). Emergence, closure and inter-level causation in biological systems. *Erkenntnis, 78*(2), 153–178.
15. Mossio, M., & Moreno, A. (2010). Organisational closure in biological organisms. *History and Philosophy of the Life Sciences, 32*(2–3), 26–288.
16. Nunes-Neto, N., Moreno, A., & El-Hani, C. (2014). Function in ecology: An organizational approach. *Biology & Philosophy, 29*, 123–141.
17. Pattee, H. H. (1973), The physical basis and origin of hierarchical control. In H. H. Pattee (Ed.), *Hierarchy theory* (pp. 71–108). Braziller: New York.
18. Piaget, J. (1967). *Biologie et connaissance*. Paris: Gallimard.
19. Rosen, R. (1972). Some relational cell models: The metabolism-repair systems. In R. Rosen (Ed.), *Foundations of mathematical biology* (Vol. II, pp. 217–253). New York: Academic Press.
20. Rosen, R. (1986). Some comments on systems and system theory. *International Journal of General Systems, 13*(1), 1–3.
21. Rosen, R. (1991). *Life itself*. New York: Columbia University Press.
22. Umerez, J., & Mossio, M. (2013). Constraint. In W. Dubitzky, O. Wolkenhauer, K.-H. Cho, & H. Yokota (Eds.), *Encyclopedia of systems biology* (pp. 490–493). New York: Springer.

Chapter 22
General System(s) Theory 2.0: A Brief Outline

Gianfranco Minati

1 Introduction

Following Bertalanffy's (1901–1972) introduction of General System Theory (GST), also known as classical Systemics [4] when intended as a corpus of concepts, approaches, or even of general social usage, the concept of system has been elaborated upon in almost all disciplinary fields, allowing inter-disciplinary approaches, including Biology, Chemistry, Cognitive Science, Economics, Education, Medicine, Physics, and Sociology.

GST, considered as General Systems Theory led to, for instance, approaches and theories such as Automata Theory, Catastrophe theory, Chaos Theory, Control theory, Cybernetics, Dissipative Structures, Games Theory, System Dynamics, and the Theory of Dynamical Systems.

2 Examples of Words and Concepts of Bertalanffy's Systemics

Examples of words and concepts used in the approaches mentioned above.

1. Anticipation;
2. Automation;
3. Completeness;
4. Computable uncertainty;
5. Context-independence;

G. Minati (✉)
Italian Systems Society, Via Pellegrino Rossi 42, Milan I-20161, Italy
e-mail: gianfranco.minati@airs.it

G. Minati et al. (eds.), *Towards a Post-Bertalanffy Systemics*, Contemporary Systems Thinking, DOI 10.1007/978-3-319-24391-7_22

6. Control;
7. Decision;
8. Degrees of freedom;
9. Forecast;
10. Growth;
11. Objective;
12. Openness;
13. Optimisation;
14. Organisation;
15. Planning;
16. Precision;
17. Regulation;
18. Reversibility;
19. Separability;
20. Solvability;
21. Symbolic;
22. Standardisation;
23. Unconnectedness.

3 Examples of Words and Concepts of Post-Bertalanffy Systemics

However, following this very fecund period, new systemic approaches, concepts, and theories arose within various specific disciplines and GST also had to deal with new problems of systemics including processes of emergence and the related acquisition of new properties [16], as well as structural dynamics as occurring in phase transitions.

Below, some examples of properties which can not be dealt with using classical Systemics:

1. Coherence as the dynamic establishment of and maintaining of a property, e.g., behavioral, shape, or topological properties, established and continuously maintained through component interactions;
2. Context-dependence, i.e., non-separability from the environment;
3. Development versus growth;
4. Emergence as continuous, irregular and unpredictable but also coherent acquisition of properties such as shape and behavior;
5. Entanglement;
6. Equivalence/non-equivalence;
7. Incompleteness as a theoretical impossibility to complete or as a degree of freedom;
8. Induction of properties rather than prescriptions or solutions;
9. Irreversibility as the price of uniqueness;
10. Low-energy effects able to break equivalences or symmetries;

11. Multiplicity, including multiple systems established by the same components;
12. Networks formed by links among properties, i.e., linked properties;
13. Non-symbolic computability given, for instance, by neural networks and cellular automata;
14. Uncertainty including uncertainty principles;
15. Non-causality, non-linear but due to networked, systems of events;
16. Non-invasiveness (see Appendix);
17. Non-prescribability but possibly only inducibility;
18. Non-separability as when expressing the observed in terms of the observer;
19. Non-symbolic representations, as for neural networks;
20. Quasi- as dynamically irregular or partial, as in quasi-periodicity;
21. Role of individuality able to break equivalences;
22. Self-Organization with continuous but stable, for instance, periodic, quasi-periodic variability in the acquisition of new structures, as for Bènard rolls;
23. Simultaneity;
24. Fractality;
25. Symmetry;
26. Structural Dynamics, changes in structure over time;
27. Uniqueness, corresponding to irreversibility;
28. Usage of degrees of freedom (see Appendix).

Extensions and updates of GST concepts are not sufficient to maintain a unitary systemic theoretical framework because of the different nature of new properties and problems which require new theoretical unitary understanding.

Examples of suitable approaches are those introduced by Network Science, Meta-structures, scale-invariance, power laws, and all Quantum-based discoveries and variations as considered below.

4 General Examples of Sources of Complexity Requiring Post-Bertalanffy Approaches

While some problems have been and can be suitably treated using Bertalanffy's GST, see above list 1.1–1.23, others cannot, see list 2.1–2.28, because of their different natures. Examples of the second kind of problems occur when studying [25]:

1. Quantum effects shown, for instance, by superconductivity;
2. Nano-technologies for chemistry, pharmacology and the science of materials;
3. Problems ignored by classical approaches such as long-range correlations; biological phenomena such as the ability of the human brain to perform enormous amounts of computations in a short time using little energy; phase transitions in cognitive problems;
4. Collective systems given by the collective motion of:

- living systems provided with sufficiently complex cognitive systems such as flocks, swarms, anthills, herds, schools, crowds, and traffic where interactions between the components involve cognitive processing;
- living systems provided with no cognitive systems such as amoeba, bacterial colonies, cells, protein chains and macromolecules;
- non-living systems such as lasers, systems of boats, nano-swimmers, nematic fluids, networks, traffic signaling, rods on vibrating surfaces, shaken metallic rods—interaction involves reacting—and simple robots where interaction occurs through simple artificial cognitive systems.

5. Collective artificial systems given by collective interactions of various natures other than physical motion include communities of mobile phone networks, industrial districts, markets, morphological properties of cities and urban development, networks such as the Internet, and queues. Interaction is given by systems with cognitive processing and reactions [20].

5 Examples of Sources of Complexity for Social Systems

Social systems are able to acquire new properties thanks to increasing knowledge; processes of emergence are activated by new configurations and self-designed structural constraints given, for instance, by architecture, cultural rules and procedures. Consider the following examples:

1. knowledge-intensiveness;
2. delocalisation and globalisation;
3. duplicability;
4. highly general networked interconnections;
5. high manipulability;
6. high virtuality;
7. hyper connections;
8. importance of individuality;
9. coherences rather than equilibrium;
10. interchangeability;
11. on-line actions;
12. reduced time between design, implementation, and marketing;
13. short general lifespan;
14. technological innovations and solutions creating new problems;
15. epiphenomena, i.e., secondary phenomena occurring alongside or in parallel to the primary phenomenon;
16. multiplicity;
17. non-linearity;
18. non-sustainability;
19. augmented reality through simulations and multi-dimensional, simultaneous, and coherent information;

20. data availability (so-called Big Data);
21. networked availability of knowledge;
22. products and services come with induction for use more than directions for use;
23. rapid transformation of solutions into new problems.

6 Changing the Concept of System?

At this point we must consider that not only new kinds of systemic properties, as listed above, are being studied in various disciplines but that the concept of system itself is under examination. This occurs when the problems and properties of systemic nature are represented, for instance, as related to networks. This is the well-known case of Network Science (see, for instance, [1, 3, 9–11, 19, 28]) where phenomena and problems are represented as networks (interdisciplinarity) and systemic properties are the properties of such networks (transdisciplinarity).

Emergent systemic properties are considered to be represented by properties of networks. The selection of nodes and links, i.e., what is to be considered as such; the kind of networks, such as scale-free, random and hierarchical; their properties, topological, scale-freeness, small worldness; modularity as a measure of the structure of networks, when detecting community structures in networks, should be considered as *a new way to represent systemic properties*.

Because of this, Network Science is often considered as the new post-Bertalanffy systems science even though other approaches should be considered, such as (a) quantum systems, which allow a number of different representations, critical interaction with the environment, and dissipation [5, 6, 8, 12, 21], when considering a possible quantum Systemics and (b) meta-structures [16, 18]. Other cases occur for collective behaviors when considering scale invariance and power laws [23], entropy [24], and topological distances [22]. When considering the DYnamical uSage of Models (DYSAM, [17, pp. 64–85]) we should look for multiple, eventually subsequent, simultaneous, and superimposed approaches to consider and model systemic properties. A new form of reductionism would be to consider one approach as unique and universal.

7 Conclusions

The first part introduced some specific aspects and concepts of classical Bertalanffy Systemics, whereas the second listed some specific aspects and concepts related to levels of complexity which cannot be approached using classical Systemics or its extensions since they are of a different nature. The third part listed examples of problems and sources possessing and generating such new situations. The fourth and last part listed examples of sources and generators of such new situations with particular reference to social systems.

We conclude by stressing that, on the one hand, the new Systemics 2.0 is not an extension of classical Systemics, but that they are expected to coexist with classical Systemics, the approaches used being possible combinations of the two Systemics at different levels. This corresponds to the fact that phenomena should be intended as combinations of functioning and emergence, and of various possible, even superimposed, representations as considered by the DYnamic uSAge of Models (DYSAM) based on strategies using independent and irreducible models as applied in well-established approaches such as the Bayesian method, Ensemble Learning, Evolutionary Game Theory, Machine Learning, Second-Order Cybernetics, and Pierce's abduction [13], [17, pp. 64–75]. There will be possible usages of one Systemics by the other.

There are overlapping fuzzy areas between the two Systemics, which are not comparable and neither is their systemic power. Approaches belonging conceptually to the former Systemics could be used to model complex phenomena, such as strange attractors for chaotic phenomena.

We conclude by stressing that the usage of Bertalanffy Systemics to deal with problems and properties of complex phenomena is to be considered as a new form of reductionism.

Appendix

This appendix presents three examples of cases using the new approaches considered by Systemics 2.0.

Orders to Complex Systems?

It is ineffective to give symbolic, explicit orders to sub-symbolic, non-explicit and emergent systems since they cannot process them. The symbolic and emergent natures are contradictory.

Possible actions cannot be explicitly prescribed but submitted in such a way as to allow their being suitably processed.

For instance, actions upon various properties of collective behaviors may relate to the environment such as using perturbations, inserting possibly dynamic obstacles, inserting perturbative phenomena such as other collective behaviors; actions upon communication between elements by using language, symbols and technologies as well as to the energy available. Orders should be translated into suggestions, contextual changes, information, and interactions to be suitably processed by the system as when using the assumption of non-invasiveness or using low energy approaches.

Non-invasiveness and Low Energy

Non-invasiveness is suggested by the limited strategic value of explicitly prescribing, or administering behavior for establishing states as mentioned above. The use of soft, low-energy approaches is very important as they do not require system interventions or the administration of intense sources of energy assuming it to be processed as through communicating vessels (like feeding a parking meter with a 10-cent coin and not with a 50-Euro banknote ...).

The assumption is that feeding the maximum is always an optimum strategy leaving the system to dose the amounts required.

However, in the latter case, the overloaded system can not explore equivalent spaces of states and trajectories from which to choose on the basis of low-energy fluctuations and influences of any kind. This relates to fundamental research in theoretical biology, for instance, by Erwin Bauer (1890–1938) who considered living systems as being different from physical ones because they do not consume the supplied energy immediately, being able to manage it.

Life would be inextricably involved in the succession and maintaining of coherence between processes of emergence and changes between the coherences of various possible states of equilibrium. One searches for coherence rather than equilibrium, coherence between multiple and dynamic equilibria, levels of coherence.

Examples exist in medicine where combinations of invasiveness and non-invasiveness occur. Any invasive interventions such as the insertion of devices, e.g., pacemakers, prostheses, transplanted organs, or removal, e.g., damaged or infected organs, should be processed, i.e., accepted and even recognized by the living system. Levels of pharmacological treatment combine different degrees of invasiveness and non-invasiveness up to the extremes of psychological treatment, as in the placebo effect [7].

Between Degrees of Freedom

Another non-invasive approach used to influence systems is the setting of suitable variable degrees of freedom to the behavior of interacting agents.

However, a more interesting aspect appears when considering not only degrees of freedom as for mechanical devices and procedures, but how they are used. For instance, while respecting the degrees of freedom, a number of generic agents use them at (a) the maximum or (b) minimum levels or (c) in regular oscillating or completely random ways [14]. The degrees of freedom DF, say $DF = [Max\,allowed - Min\,allowed]$ may be used at different percentages and may adopt different temporal sequences. Such temporal sequences should be intended as behavioral profiles, irrelevant at macroscopic levels but significant at microscopic levels and in determining mesoscopic properties.

Another example is given by the structuring of space in which agents interact, as in urban planning when deciding the shape and size of roads, the inclusion of roundabouts and speed bumps in order to influence the properties of traffic, crowd evacuation in case of emergency or managing long queues; rooms in schools, hospitals and offices [2, 15, 26, 27, 29].

References

1. Barabási, A. L. (2002). *Linked: The new science of networks*. Cambridge, MA: Peruses Publishing.
2. Batty, M. (2013). *The new science of cities*. Cambridge, MA: MIT.
3. Bauke, H., Moore C., Rouquier, J.-B., & Sherrington, D. (2011). Topological phase transition in a network model with preferential attachment and node removal. *European Physical Journal B, 83*, 519–524.
4. Bertalanffy, L. V. (1968). *General system theory. Development, applications*. New York: George Braziller.
5. Blasone, M., Jizba, P., & Vitiello, G. (2011). *Quantum field theory and its macroscopic manifestation*. London: Imperial College Press.
6. Breuer, H.-P., & Petruccione, F. (2007). *The theory of open quantum systems*. Oxford: Oxford University Press.
7. Colloca, L., & Benedetti, F. (2005). Placebos and painkillers: Is mind as real as matter? *Nature Reviews Neuroscience, 6*, 545–552.
8. Davies, E. B. (1976). *Quantum theory of open systems*. London: Academic Press.
9. Derényi, I., Farkas, I., Palla, G., & Vicsek, T. (2004). Topological phase transitions of random networks. *Physica A, 334*, 583–590.
10. Dorogovtsev, S. N., Goltsev, A. V., & Mendes, J. F. F. (2008). Critical phenomena in complex networks. *Reviews of Modern Physics, 80*, 1275–1335.
11. Lewis, T. G. (2009). *Network science: Theory and applications*. Hoboken, NJ: Wiley.
12. Matsumoto, H., Sodano, P., & Umezawa, H. (1979). Extended objects in quantum systems and soliton solutions. *Physical Review D, 19*, 511–516.
13. Minati, G. (2007). Some new theoretical issues in systems thinking relevant for modelling corporate learning. *The Learning Organization, 14*(6), 480–488.
14. Minati, G. (2012). Introduction to the meta-structures project: Prospective applications. *World Futures, 68*(8), 558–574.
15. Minati, G., & Collen, A. (2009). Architecture as the cybernetic self-design of boundary conditions for emergent properties in human social systems. *Cybernetics & Human Knowing, 16*(1–2), 101–123.
16. Minati, G., & Licata, I. (2012). Meta-structural properties in collective behaviours. *The International Journal of General Systems, 41*(3), 289–311.
17. Minati, G., & Pessa, E. (2006). *Collective beings*. New York: Springer.
18. Pessa, E. (2012). On models of emergent metastructures. In G. Minati, M. Abram, & E. Pessa (Eds.), *Methods, models, simulations and approaches towards a general theory of change* (pp. 113–134). Singapore: World Scientific.
19. Valente, T. W. (2012). Network interventions. *Science, 337*(6090), 49–53.
20. Vicsek, T., & Zafeiris, A. (2012). Collective motion. *Physics Reports, 517*, 71–140.
21. Weiss, U. (2012). *Quantum dissipative systems* (4th ed.). Singapore: World Scientific.

Web Resources

22. Ballarini, M., Cabibbo, N., Candelier, R., Cavagna, A., Cisbani, E., Giardina, I., et al. (2008). Interaction ruling animal collective behavior depends on topological rather than metric distance: Evidence from a field study. *Proceeding of the National Academy of Sciences of the*

United States of America, 105(4), 1232–1237. Accessed November 2013. Available from http://www.pnas.org/content/105/4/1232.full.pdf+html

23. Cavagna, A., Cimarelli, A., Giardina, I., Parisi, G., Santagati, R., Stefanini, F., et al. (2010). Scale-free correlations in starling flocks. *Proceeding of the National Academy of Sciences of the United States of America, 107*, 11865–11870. Accessed November 2013. Available from http://www.pnas.org/content/early/2010/06/11/1005766107.full.pdf+html
24. Cavagna, A., Giardina, I., Ginelli, F., Mora, T., Piovani, D., Tavarone, R., et al. (2013). Dynamical maximum entropy approach to flocking, theory (on line). ArXiv. Accessed November 2013. Available from http://arxiv.org/abs/1310.3810
25. Minati, G., & Licata, I. (2013). Emergence as mesoscopic coherence. *Systems, 1*(4), 50–65. Accessed November 2013. Available from http://www.mdpi.com/2079-8954/1/4/50
26. Space Syntax Laboratory. Accessed November 2013. Available from http://www.spacesyntax.org/
27. The Behavioral Design Lab. Accessed November 2013. Available from http://www.designcouncil.org.uk/our-work/insight/behavioural-design-lab/
28. The Center for Complex Network Research. Accessed November 2013. Available from http://www.barabasilab.com/
29. University College London (UCL) Centre for Advanced Spatial Analysis (CASA). Accessed November 2013. Available from http://www.casa.ucl.ac.uk/

Chapter 23
Phenomenology of Emergence in Music: Presentation of the Processes of Systemic Emergence in the Contrapuntal and Improvizational Aspects of Baroque Music

Emanuela Pietrocini

1 Phenomenology

To appearance we owe concepts that let themselves be thought for themselves.[1]

This is what Johann Heinrich Lambert, German philosopher and mathematician from the eighteenth century, affirms concerning the manifestation and representation of "simple concepts"[2] through appearance which, although not containing its foundation, is sufficient to represent the ideal and necessary bridge between sensory perception and intellect; therefore, the solid structures of experience do not lay on the level of empirical propositions but on that of "universal possibilities" [29].

The link between the thinkable and perception is at the basis of Lambert's *Phänomenologie*; he was the first to use such term to define the critical science of perception, the instrument with which to trace convergences and divergences between absolute idea and sensory experience. Lambert's position, within the context of the reflection on the sciences, lies between Leibniz and Kant in postulating, on the one hand, the validity of the mathematical method in metaphysics; on the

E. Pietrocini (✉)
Early Music Departement Schola Palatina, La Vertuosa Compagnia de' Musici di Roma, ISIA of Pescara, Multimedia Design, Pescara, Italy
e-mail: dipartimentomusica@gmail.com

[1] "Wir haben ferner dem Schein Begriffe zu danken, die sich für sich gedenken lassen". In J.H. Lambert, Neues Organon, oder Gedanken über die Erforschung und Bezeichnung des Wahren und dessen Unterscheidung von Irrtum und Schein, Teil 4: Phänomenologie oder Lehre vom Schein, J. Wendler, Leipzig, 1764, p. 248, §53.

[2] "Simple concepts", in Lambert's conception, are the primary elements of knowledge: they are pure content and are recognizable through the perception of sensory manifestations. Even though they cannot be formalized, simple concepts can be described as absolute generalizations (for example axioms and postulates), a quality which makes them both gnoseological foundations and material principle of knowledge.

G. Minati et al. (eds.), *Towards a Post-Bertalanffy Systemics*, Contemporary Systems Thinking, DOI 10.1007/978-3-319-24391-7_23

other hand, the necessity to attribute the flux of science to the *phenomenon* [14]. The legacy of scientific rationalism, therefore, moves to the age of Enlightenment, especially through Locke, Wolff and Leibniz; it is actually thanks to the latter that we owe the reconciliation between senses and reason which allows the "music that can be heard"—namely the music which manifests itself to sensory perception—to acquire the dignity of science and to become at the same time the object of physical survey and of metaphysical reflection. In the Pythagorean-Platonic conception and, later, in Boethius and in the medieval conception, only *musica mundana*—celestial harmony and music of the spheres—can be considered as a science [13]; it is not available to any sound manifestation but it represents the harmonious movement of the cosmos through the Law of the Number.

Only with Gioseffo Zarlino, in the full bloom of the Renaissance, the metaphysical concept of *musica mundana* is re-interpreted in a secular and Humanistic way:

> mundane harmony is not only among the things that can be seen and that are known in the heavens; it is also in the connection of the Elements and in the variety of tempos that may be understood.[3]

The aim is to detect an immanent rational principle in the relationships among sounds and this is indeed the research field developed by Zarlino through his studies on harmonic sounds, on natural modes, on the division of the octave and temperament. The passage of music to scientific dignity—although fundamental—refers exclusively to the study of physical and acoustic matters and to theoretical issues, even though connected to the musical practice. It is during the Baroque period that the attention focuses, at last, on musical composition, on the structures and on the laws which regulate it, on the connections that composition creates with—and among—the fields of knowledge.

Once again, it was philosophers and mathematicians such as Keplero,[4] Mersenne[5] and Descartes[6] who dealt with music, substantially maintaining the dualistic position between the music of the senses and the music of intellect, as well as exceptional and eclectic personalities such as Athanasius Kircher. The Jesuit, defined as "master in about a hundred arts" (Woods 2005, p. 108) is among the most famous intellectuals of his time: his scientific researches embrace any possible field of knowledge, from Egyptology to Geology, from the observations on micro-organisms and on infectious diseases (he was among the first to use the microscope and to formulate

---

[3] "La mondana è quell'armonia che non solo si conosce essere tra quelle cose che si veggono e si conoscono nel cielo: ma nel legamento de gli Elementi et nella varietà dei tempi ancora si comprende". In LE ISTITUTIONI/ HARMONICHE/ DI M. GIOSEFFO ZARLINO DA CHIOGGIA;/Nelle quali; oltre alle materie appartenenti/ALLA MUSICA;/si trovano dichiarati molti luoghi/di Poeti, d'Historici, & di Filosofi/.../in Venetia MDLVIII.

[4] Ioannis Keppleri/ HARMONICES/MUNDI/LIBRI V/Lincii Austriae/Sumptibus GODOFREDI TAMBACHII Bibl. Francof./Excudebat IOANNES PLANCUS./ANNO MDCXIX.

[5] HARMONIE/VNIVERSELLE,/CONTENANT LA THEORIE/ET LA PRATIQUE/DE LA MUSIQVE/ .../Par F. Marin Mersenne de l'Ordre des Minimes/A PARIS,/Chez Sebastien Cramoisy, Imprimeur ordinaire du Roy,/ruë S. Iaques, aux Cicognes/MDCXXXVI.

[6] RENATI/DES-CARTES, MUSICÆ/COMPENDIUM./TRAJECTI AD RHENUM,/TYPIS Gisberti à Zyll, & Theodori ab Ackersdyck,/MDCL.

valuable hypothesis on the origin of plague), to mechanical and technological invention. Among his many works, *Musurgia Universalis*,[7] published in 1650 in two volumes and ten books, aims at creating a systematic exposition of the theoretical and practical foundations of music without neglecting any aspect or implication [21]. Extremely far from the tendency to specialization which originates in his times, Kircher strongly advocates the uniformity of knowledge, affirming that the universal order of Nature manifests in the same way in all fields of knowledge.

The mathematical medium is applied in a distinguishing and pervasive way in the eighth book, *de Musurgia Mirifica*: the author presents his own compositional method based on combinatorial computation, through which anyone, even those totally unfamiliar with the most elementary musical notions, could try and compose a contrapunctal vocal piece for four voices [5].

Kircher's method is naturally part of the musical theory and practice of the time: seventeenth century counterpoint is regulated by interval relations which can be expressed numerically, whose concatenations and successions match with mathematically defined structures; Harmony itself follows analogous procedures, redefined in temporal verticality [9]: the *Basso Continuo*[8] is codified using numbers. The model proposed by the Jesuit tends to prefer the means rather than the result, pretending to attribute each aspect of the composition—the expressive and semantic aspect included—to the sole use of combinatorial algorithms, thus neglecting—completely and consciously—the aesthetic component, the qualitative judgement and the faculty of expressive choice which are inherent to artistic creation.

2 Primary Simplicities

Baroque music makes use of both the contrapunctal and the harmonic procedures, according to the formal reference points, contexts, languages and uses the composition is destined to; in both procedures, the techniques of re-processing of the sound material essentially based on combinatorics, on topological transformation and on mutation are used. Generally speaking, the contrapunctal compositions are entirely noted and the score appears complete in all parts; the production based on the harmonic development of a main melody accompanied by a bass (*Cantate*, *Arie*, *Sonate*, *Suites* of dances) is only partially noted because the mentioned

7 ATHANASII KIRCHERI/FULDENSIS E SOC JESU PRESBYTERI/MUSURGIA/UNIVERSALIS/SIVE/ARS MAGNA/CONSONI ET DISSONI/IN X LIBROS DIGESTA./Qua Vniversa Sonorum doctrina,/et Phylosophia, Musicaeque tam theoricae,/quam practicae scientia, summa varietate traditur; admirandae Consoni, et Dissoni in mundo, adeoque/Universa Natura vires effectusque, vti noua, ita peregrina variorum speciminum/exhibitione ad singulares usus, tum in omnipoene facultate, tum potissimum/in Philologia, Mathematica, Physica, Mechanica, Medecina, Politica,/Metaphysica, Theologia aperiuntur et demonstrantur./ Tomus I/ROMAE, Ex Typographia Haeredum Francisci Corbelletti, Anno Iubilaei MDCL; Tomus II/ROMAE, Typis Ludouici Grignani, Anno Iubilaei MDCL.

8 Thorough Bass: extempore realization of a harmonic-contrapunctal accompaniment of a main melodic line, supported by a given bass [10].

development, which includes the realization of the chords of accompaniment, the imitation, the re-processing of the same melody and the ornamentation, are left to the extemporization of the performance. Throughout the whole of the seventeenth century, the vocal and instrumental production of accompanied monody makes extensive use of forms based on cyclic structures and *ostinati* like *Ciaccone*, *Passacagli* and *Ground*, or of variations on a given bass (*Partite*), generally taken from melodies of popular origin then transferred to the cultured tradition and proposed as models (the *Folia*, the *Monicha*, the *Romanesca*, *Ruggiero* etc. [3]). Such melodies offer a wider creative space as well as the best opportunities for expression; they serve the re-processing admirably because they are constructed in the most essential and in the simplest way, according to the natural principles of relations and proportions:

> Music fascinates us even if its beauty consists only in the consistency of numbers and of computation—computation of which we are not aware and which the soul does constantly—in the dissonances or vibrations of the sound bodies which meet in some intervals.[9]

Lambert's issue concerning simple concepts is also found in music. What can be defined as a primary element among the perceptions of the musical phenomenon? Certainly the sound itself, manifesting through the senses; however, it is defined *exemplarisch gelernt*, an exemplary learning, because it contains the absolute possibility of simple concepts, but cannot represent them. That's why the mind needs *Übung*, exercise. A whole of practices and exercises replaces verbal definitions and allows an intuitive introduction of such primary simplicities [29, p. 159].

What comes to mind is the cyclic formulas which in music are realized with the use of only one interval; it can be the octave or the fifth, and the sounds deriving in fact exist and are one sound only, according to natural harmonics. This type of relation is therefore contained in the existence of sound itself, it is perceived *sub specie senso* and historically displayed in the construction of cyclic models based on alternation, the same that Diego Ortiz, in his "Trattado de Glosas" appeared in 1553 defines

> plain chants which are generally called Tenores in Italy.[10]

The already mentioned *Romanesca* appears in the second book of the "Trattado", dedicated to

> … the different manners to be played with the violone together with the harpsichord, ….[11]

---

[9] "La Musique nous charme quoique sa beauté ne consiste que dans les convenances des nombres et dans les compte, dont nous ne nous apercevons pas, et que l'ame ne laisse pas de faire, des battements ou vibrations des corps sonnans, qui se rencontrent par certains intervalles". G.W. Leibniz, Principes de la nature et de la grace fondés en raison (1714); in God. Guil. Leibnitii/Opera Philosophica/Quae Exstant/Latina, Gallica, Germanica/Omnia/Edita Recognovit E Temporum Rationibus Disposita Pluribus Ineditis Auxit/Introductione Critica Atque Indicibus/instruxit/Joannes Eduardus Erdmann/pars altera/ Berolini./ sumtibus G. Eichleri/MDCCCXXXIX.

[10] "… cantos llanos que en Italia comunemente llaman Tenores … ". In DE DIEGO/ORTIZ/ TOLEDANO/LIBRO SECONDO in TRATTADO/de Glosas sobre/Clausulas y otros/generos de puntos/en la Musica de/Violones nueva=/mente puestos/en luz./ En Roma por Valerio Dorico, y Luis/su hermano a x. de Dezemb./1553, p. 47.

[11] "… le varie maniere che si debbiano sonare col Violone, e col cimbalo insieme … ", Ibid. p. 26.

Fig. 23.1: Romanesca bass, from D. Ortiz, cit. p. 58

Among the exempla illustrating the second manner of playing,[12] namely the improvisation on a given chant,

> for a better completion of such work.[13]

The author presents the *Romanesca* in the polyphonic version, in four parts, recommending to keep the attention focused on the *baxo solo* for the processing of the *Recercada*.[14] In fact, the basic elementary structure of the piece is the bass. It is made of a very brief series, laid out according to the relations of fifth as shown in Fig. 23.1.

Then follow two magnificent *Recercadas* for viola da gamba (or Violone), offered as examples of improvisation like exercises to be practiced so that

> ... every student can improve following a good order and playing reasonably and not randomly.[15]

Here is the exercise, through which the intellect learns to recognize the elementary principles of knowledge in order to represent them and to re-arrange them within a net of relations based on number and proportion. The connection with Leibniz' conception, according to which the soul makes an unconscious calculation in making use of Music, representing in such calculation the utmost pleasure for itself and for the intellect, appears evident here: in fact, the improvisation on a given chant responds to regulations and distinctive signs that are typical of a language and of a system of relations which is mathematically well defined [17]. The Baroque, as already mentioned, witnesses the consideration on music move to the scientific field [4]; we may even imagine that the specific characteristics of the path of development which occurs in seventeenth century music in Europe—really extraordinary in terms of dynamicity and multiplicity—have contributed to denote an ontological necessity destined to change the approach to knowledge forever.

[12] "... seconda maniera di sonare ... ", Ibid. p. 26.

[13] "... para major cumplimiento desta obra ... ", Ibid. p. 47.

[14] Type of improvisational piece on a given chant.

[15] "... qualunch'studioso potersi con un bel ordine procedere e sonar per ragione e non a caso ... ", D. Ortiz, op. cit., p. 47.

3 Science, Knowledge and Music

The path followed by Leibniz—who stumbled upon the art of sounds since his juvenile studies—is emblematic. It was an unavoidable contact for a student determined to examine mathematics, philosophy and theology in depth, in a cultural context imbued with neopythagoreanism and neoplatonism. Initially, he conceives music only in its indissoluble and metaphysical relation with the Number; what intrigues him the most is in fact the relations among numbers; he thought that reality is composed of few primal elements which, combining and intermingling among them in different ways, give life to all that exists [31]. Music functions in a similar way: a system of only twelve sounds can give life, through different systems of relation, to a virtually infinite number of musical compositions of the most diverse types.

Music as a metaphor of the universal mechanism is a recurrent subject in Leibniz's researches and, in his work, nearly acquires a de-stabilizing function; each time, the philosopher detects a method of survey and an aspect of the research which he follows with utmost coherence and determination, even for long years, until something indefinite, although intensely perceived as an inconsistency, a subtle contradiction, urges him towards a different direction.

The fascination for the monumental work by Kircher on music inspires one of his first works, the *Dissertatio de Arte Combinatoria*, written and edited in 1666, in which young Leibniz proposes to develop a method of mathematical calculation [26] which allows to explain scientifically any aspect of human speculation, so that any barrier among different disciplines can be eliminated and all disciplines can be unified under the aegis of its unquestionable authority [31, p. 16]. The references to music can be detected in plenty, in particular concerning the representation of the huge potentials of combinatorics applied to mechanical engineering, to the construction of complex musical instruments like the pipe organ [31, p. 20], and to the same musical composition. Leibniz's enthusiasm for combinatorics will be further increased by the publication of *Ars magna sciendi*,[16] a work which Kircher dedicates exclusively to the calculation of combinations; the young philosopher finally decides to write to his Mentor, expressing all his admiration and acknowledging the lineage of his own work, which he submits to the master's evaluation, to the *Musurgia* by the great Jesuit.

In his letter dated 16th May 1670, Leibniz refers to the famous *Arca Musarithmica*, the futuristic mechanical instrument designed by Kircher in order to automate the process of formation of the melodic-rhythmical lines that can be associated to the text to be put to music; he asks for further information, he entreats him so that

[16] ATHANASII KIRCHERI,/E Soc Jesu/ARS MAGNA/SCIENDI/in XII Libros Digesta,/QUA/NOVA & UNIVERSALI METHODO/Per Artificiosum Combinationum contextum de omni/re proposita plurimis et prope infinitis rationibus disputari omniumque/summaria quaedam cognitio comparari potest./AD/Augustissimum Rom. Imperatorem/LEOPOLDUM PRIMUM,/Justum, Pium, Felicem, /Amstelodami, Apud JOANNEM JANSSONIUM À WAESBERGE,/& viduam Elizei Weyerstraet,/ Anno MDCLXIX.

he can be allowed to study in depth the most occult and complex aspects of the *Ars Magna*, of which he senses tacit implications. Kircher replies coolly, skirting the issues which by the way are expressed rather vaguely and indefinitely in his own works. *Musurgia Mirifica*, the extraordinary art of composing according to the sole laws of combinatorics [30], does not reveal the intimate connections between the world of sounds and that of feelings and emotions. The references to the doctrine of the affections and the hints to "*musurgia rethorica*" converge, ultimately, to a question: such knowledge, of the utmost and exclusive importance, is reserved to the few chosen ones who showed to be worth receiving it. Nobody was ever given that privilege by the Jesuit.

From this moment, Leibniz's attention to music and combinatorics fades away, and focuses, years later, on the researches on acoustics and musical theory which he will share at length with mathematician Conrad Henfling. Such new field of survey will lead him to confront—among other things—with the thorny problem of Temperament,[17] subject to plenty of speculations and controversies which see him deeply involved together with the greatest scientific personalities of his time, like Huygens and Sauveur. However, even this channel of research will prove insufficient; after about 6 years, the intense epistolary exchange with Hefling is definitely interrupted. It stops, about 2 years earlier, with a consideration fully revealing the distance taken from the theoretical speculation on music:

> ... I should hope that we would think more than we normally do about the reasons of the practice and of what is most liked in the compositions ...[18] [15, pp. 146–147].

The philosopher takes his definite distance from the mathematical method to take the direction of the theorization of what we may define an ontological aesthetics of art [27]. The perception of sensory phenomena occurs in a confuse and indistinct way; when such phenomena imply—like sound and music—numeric relations, structures and proportions, the soul activates in the exercise of unconscious calculation, thus inducing, at the spirit level, the emotional reaction. The numbers composing intervals make it possible for consonances and dissonances to be represented in the soul according to the same principle, producing either delight or unpleasantness; the harmonic perception of consonant intervals, like the octave and the fifth, manifests as a vague and ineffable feeling of pleasure—what Leibniz calls *je n'sais quoi* [6].

[17] Temperament consists of the artificial determination of sounds and intervals within the octave, due to the special needs that instruments with fixed tuning have. Since the natural definition of pitch obtained by natural harmonics or by the succession of perfect fifths determines a residual interval, called Pythagorean comma, the system does not allow the transposition to the different tones, therefore in time, one adopts different empirical or mathematical methods for the distribution of such difference between the intervals.

[18] "... je souhaiterois qu'on pensat un peu plus qu'on ne fait ordinairement, aux raisons de la practique et de ce qui plaist le plus dans les compositions ... " Leibniz to Conrad Henfling, April 1709.

4 Fifths, Octaves and Emergence

Here we are back to our *Romanesca* and to the interval of fifth. According to Leibniz's aesthetics we can easily understand the reason for the enormous popularity that it enjoyed throughout the whole production of Baroque music: virtually all the greatest composers of the time, over a century and a half, elected it as the foundation for all sorts of re-processing, from improvisational forms based on variation to strict counterpoint [16]. Concerning the first, we have numerous examples, both in vocal and instrumental literature, since the forms and the processes typical of extemporization have been gradually assimilated by the compositional techniques, determining on the one hand the movement of the space-time dimensionality from the continuum of impermanence to the discretion of codification [1]; on the other hand the reflexive relocation of the processes of mutation. We are not dealing with a sort of "freeze-frame", like the reduction of a film to single frames, but with the representation of a perceptive experience presenting itself to the intellect by induction and then transforming itself into consciousness [25]. We find in these compositions all the characteristics that, from a systemic point of view, detect the processes of Emergence: non-linearity, unpredictability, change, adaptability, non-definition of procedural limits [22]. The use of destabilizing elements, like unprepared and unresolved dissonances, relocation of tonic accents and unaligned rhythmical solutions, is presented in a pervasive and structural way, but, instead of manifesting as artificial, unnatural and disagreeable, it determines a global harmonic effect perceived by the observer-listener as totally exempt from fractures or inconsistencies. What in improvisation presents itself spontaneously is what we could define as a process typical of complexity: relations between elements and systems of elements establish themselves in the space-time continuum represented by extemporization; models and processes superimpose giving life to new forms of interaction and new structures. The phenomena of *interference*, just hinted at, present themselves as both function and product of the dynamic processes of Emergence, which activate in that particular condition of crisis that is impermanence [23]: when improvisation occurs around a generating idea, like a given chant, there is a thematic redundancy which implies the request of variation. The procedural principles of variation models refer mainly to combinatorics, used for changing formal and structural parameters of the sound organization; however, although they are potentially sufficient to satisfy unlimitedly the technical needs of composition, they also undergo a change describing strong methodological distractions from the mathematical model and—often—from any form of local stability. Therefore, we detect phenomena of Emergence both in the manifestation of unpredictable elements and inconsistent solutions within the system of reference, as well as in the consolidation of unusual procedures, far from any describable stylistic model in a linear way, superimposed or alternated to others that are utterly regular and common. It is possible to observe the evidence of such phenomenon in those pieces that, originated by improvisational forms, have been codified and represented—thus transferred to that field of the reflexive and conscious operation which is defined composing music; unexpectedly, we will be able to detect similar characteristics even in compositions that are created to be pure representations of thought, musical architectures perfectly structured and in detail according to the strictest rules of Counterpoint.

5 Coherences

As an example, without any analytical pretence, we will briefly examine some of the evidences of the manifestation of Emergence in music in two Baroque compositions, very far from each other in terms of chronology, form and style. The common element is represented by the characteristics of the subject on which they are built: in both cases it is a cyclic bass, of the type mentioned earlier, namely of a *Romanesca* in the first and of a *Ciaccona*[19] in the second.

The Madrigal "Ohimé dov'è il mio ben"[20] by Claudio Monteverdi is defined as *Romanesca à due* from the very first Venetian edition in 1619. It is a composition for two voices and *basso continuo*, in four parts built on the same bass, whose poetic text in *ottava rima* is by Bernardo Tasso; we can detect poetic formulas and rhetorical figures typical of the period, which the composer does not fail to interpret according to the consolidated rules of the *affektenlehre*[21]; the first part in fact, opens on the word *ohimé* with the introduction of the second voice on the first in a minor second: an extremely dissonant interval, justified by the sorrowful exclamation of the text. Later on, the said interval and other dissonances appear recurrently, regardless of the sense of the words and of the rhetorical figures; frequently, the dissonance appears following a rhythmical anticipation, at times it introduces it, thus determining a detachment from the harmonic structures serving the realization of the *Continuo*. In one case in particular, in the second part, the two voices present an interval of minor third, a very harmless interval, but they create with the bass respectively a minor second and a minor seventh, with a totally bewildering effect. The consequence of such *interference*, which, at first glance seems to be losing itself in a consonant resolution, is the sudden ensuing of a pacy progression, crammed with increasingly discordant intervals, which characterizes the whole pace of the section.

The second piece belongs to the *corpus* of the didactic works by Johann Sebastian Bach [8]. It is one of the 15 *Sinfonien* for three voices, written between 1722 and 1723 and addressed to students, music lovers and

[19] Musical form created on a ground bass in 3/4 tempo of improvisational nature derived from the homonymous sixteenth century dance of Spanish/Latin-American origin.

[20] CONCERTO/SETTIMO LIBRO/DE MADRIGALI/A 1. 2. 3. 4. & Sei voci, con altri/generi de Canti,/DI/CLAUDIO MONTEVERDE/MAESTRO DI CAPPELLA/ Della Serenissima Republica/Novamente Dato In Luce/DEDICATO/ALLA SERENISSIMA MADAMA/CATERINA DE MEDICI/Gonzaga Duchessa di Mantoua di Monferrato &c/STAMPA DEL GARDANO. IN VENETIA MDCXIX/ Appresso Bartolomeo Magni.

[21] *Affektenlehre*, that is, the doctrine of the affections; it consists in the application of rhetorical forms and figures to the musical language, in order to arouse emotions, feelings and specific suggestions [2].

> . . . most of all to those who are eager to learn (. . .) to make good use of three obliged voices (. . .) and to acquire the art of cantabile and the taste for the composition.[22]

The *Sinfonia* we are dealing with is the one named BWV 795, the ninth according to the ascending succession of tonalities, in F minor. The bass is composed of a descending chromatic series of six notes, two of which are passage notes; by deleting them, we obtain the basic sequence, that is, the descending tetrachord typical of the *Ciaccona*. From this minimal series derive, other than the thematic subject, a second and a third series, built through contrapunctal procedures on the principle of *diminuzione* [11], that is, of the rhythmical processing through a major number of notes to short values for each original sound. It is not possible, neither it is consistent with this dissertation, to describe extensively the compositional technique used in this piece; in the attempt to give a general definition, we will say that it consists in determining processes of compositional reprocessing through the interaction among the three series, using the thematic exchange among the voices, the topological variation of intervals and of the melodic direction, as well as permutations [12]. We could detect the triumph of combinatorics wished for by Kircher as well as by young Leibniz. The result however goes much farther than expected: Bach chooses to use to the extreme the wide margins of freedom offered by the chromatic nature of the original series [7], thus pushing himself to the limits of the audible. The whole composition is pervaded by dissonances more or less taken to resolution; among them stands out the augmented fourth interval, the ill-famed *diabolus in musica*[23] which functions as a generator of dynamicity introducing, each time it appears, sudden changes. In this case *interference*, used consciously and in a functional way, takes a structural role characterizing the process according to the emerging systemic features.

6 Conclusions

How does the product of the processes of Emergence in music manifest to the observer-listener [18]? In what way can we represent the survey of the emerging features connected to the synaesthetic fruition of the musical phenomenon? According to experience, we may affirm that perception, at the same time immanent and impermanent, of the sound object, amounts to the intuition of the beauty represented. As Leibniz and Lambert said, the soul catches number and proportion

---

[22] "Auffrichtige Anleitung,/Wormit denen Liebhabern des Clavires,/besonders aber denen Lehrbegierigen, eine deüt-/liche Art gezeiget wird, nicht alleine (1) mit 2 Stimen/reine spielen zu lernen, sondern auch bey weiteren pro-/greβen auch (2) mit dreyen obligaten Partien richtig/und wohl zu verfahren, anbey auch zugleich gute inventio-/nes nicht alleine zu bekommen, sondern auch selbige wohl/durchzuführen, am allermeisten aber eine cantable/Art im Spielen zu erlangen, und darneben einen/starcken Vorschmack von der Composition zu über-/kommen./Verfertiget" /Anno Christi 1723/von Joh: Seb: Bach./Hochfürstlich Anhalt-Cöthe-/nischen Capellmeister. Frontispiece of the autograph dated 1723 kept in the Staatsbibliothek in Berlin.

[23] Interval consisting of three tones, also known as *Tritono*, having a strongly dissonant effect.

through the *petites perceptions*, which activate the unconscious action of "counting music" and delineate the ideal representation of simple concepts, sources of all knowledge.

In systemic terms, we will say that the process of Emergence must be indissolubly linked to the existence of an observer who has built a model of a given system, introducing in it such rules and symmetries that can satisfy the general principles that are considered valid [20, pp. 89–97]. On higher levels of complexity, the role of the observer consists in noticing coherence [19]: there is no interruption between original systems and the emerging ones; what is determined is rather a superimposition, a form of coexistence that can give life, subsequently, to new interactions. The level of abstraction is undoubtedly higher but, from the point of view of phenomenology, the act of "acknowledging coherence" manifests spontaneously and almost immediately, as it occurs in the perception of balance and harmony in the performances of art [24]. The secrets jealously hidden by Kircher behind the omnipotence of *Ars Magna* may have lost their mysterious appeal, but they still remain in the very heart—void and resplendent—of *thinking music*.

References

1. Apel, W. (1962). *Die notation der polyphonen Musik 900–1600*. Leipzig: Breitkopf & Härtel Musikverlag.
2. Bartel, D. (1997). *Musica poetica. Musical-rethorical figures in German baroque music*. Lincoln: University of Nebraska Press.
3. Bianconi, L. (1982). *Il Seicento in Storia della Musica* (Vol. IV). Torino: Società Italiana di Musicologia, EDT.
4. Bukofzer, M. F. (1947). *Music in the baroque era. From monteverdi to bach*. New York: W.W. Norton & Company (Italian translation: *La musica barocca*, a cura di P. Isotta e O.P. Bertini, Rusconi, Milano, 1982).
5. Chierotti, M. (1994). Comporre senza conoscere la musica: Athanasius Kircher e la Musurgia Mirifica. Un singolare esempio di scienza musicale nell'età barocca. *La Nuova Rivista Musicale Italiana, 28*, 382–410.
6. D'Angelo, P., & Velotti, S. (1997). *Il "non so che", storia di un'idea estetica*. Palermo: Aesthetica.
7. David, H. T., & Mendel, A. (1966). *The Bach reader*. New York: W. W. Norton.
8. de Candè, R. (1984). *Jean-Sébastien Bach*. Paris: Éditions du Seuil (Italian translation: *Johann Sebastian Bach*, edited by P. Peroni, Edizioni Studio Tesi, Pordenone, 1990).
9. de la Motte, D. (1981). *Kontrapunkt – Ein Lese und Arbeitsbuch*. Kassel: Bärenreiter-Verlag (Italian translation: *Il Contrappunto, un libro da leggere e da studiare*, Ricordi & C., Milano, 1991).
10. Del Sordo, F. (1996). *Il Basso Continuo*. Padova: Armelin Musica sas – Edizioni Musicali Euganea.
11. Erig, R., & Gutmann, V. (Eds.). (1979). *Italienische diminutionen die zwischen 1553 und 1638 mehrmals bearbeiteten sätze*. Zurich: Amadeus Verlag – Bernhard Pauler.
12. Ferguson, H. (1975). *Keyboard interpretation from the 14th to the 19th Century*. New York: Oxford University Press.
13. Fubini, E. (1976). *L'estetica musicale dall'antichità al Settecento*. Torino: Einaudi.
14. Fubini, E. (1986). *Musica e cultura nel Settecento europeo*. Torino: EDT.
15. Haase, R. (Ed.). (1982). *Der Briefwechsel zwischen Leibniz und Conrad Henfling*. Frankfurt: Klostermann.

16. Howen, H. (1992). *Modal and tonal counterpoint from Josquin to Strawinskj*. Belmont, CA: Wadsworth Group/Thomson Learning (Italian translation: *Il Contrappunto modale e tonale da Josquin a Strawinskj*, Ed. Curci, Milano, 2003).
17. Mattax Moersch, C. (2009). Keyboard improvisation in the baroque period. In G. Solis & B. Nettl (Eds.), *Musical improvisation, art, education, and society* (pp. 150–170). Champaign, IL: University of Illinois Press.
18. Minati, G. (2010). Sistemi: origini, ricerca e prospettive. In L. Urbani Ulivi (Ed.), *Strutture di Mondo, il pensiero sistemico come specchio di una realtà complessa* (Vol. I, pp. 15–46). Bologna: Il Mulino.
19. Minati, G. (2013). Note di sintesi: novità, contributi, prospettive di ricerca dell'approccio sistemico. In L. Urbani Ulivi (Ed.), *Strutture di mondo, il pensiero sistemico come specchio di una realtà complessa* (Vol. II, pp. 315–336). Bologna: Il Mulino.
20. Minati, G., & Pessa, E. (2006). *Collective beings*. New York: Springer.
21. Pangrazi, T. (2009). *La Musurgia Universalis di Athanasius Kircher. Contenuti, fonti, terminologia*. Firenze: Olschki.
22. Pessa, E. (2002). What is emergence? In G. Minati & E. Pessa (Eds.), *Emergence in complex cognitive, social and biological systems* (pp. 379–382). New York: Kluwer Academic/Plenum Publishers.
23. Pietrocini, E. (2012). Music: Emergence and metastructural properties in the practice of the thorough bass. In G. Minati, E. Pessa, & M. Abram (Eds.), *Methods, models, simulations and approaches towards a general theory of change* (pp. 633–646). Singapore: World Scientific.
24. Pietrocini, E. (2013). Musica dell'Emergenza. Prospettive di ricerca sistemica in musica e in musicologia. In L. Urbani Ulivi (Ed.), *Strutture di Mondo, il pensiero sistemico come specchio di una realtà complessa* (Vol. II, pp. 289–313). Bologna: Il Mulino.
25. Pineau, M., & Tillmann, B. (2001). *Perception des structures musicales*. Paris: L'Harmattan.
26. Rossi, P. (1983). *Clavis universalis. Arti della memoria e logica combinatoria da Lullo a Leibniz*. Bologna: Il Mulino.
27. Tomasi, G. (2002). *La bellezza e la fabbrica del mondo: estetica e metafisica in G.W. Leibniz*. Pisa: ETS.
28. Woods, T. (2005). *How the catholic church built western civilization*. Washington, DC: Regnery Publishing.

Web Resources

29. Basso, P. *Filosofia e geometria. Lambert interprete di Euclide*. Collana Il Dodecaedro. Available from http://www.lettere.unimi.it/dodeca/basso/coperti.htm
30. Chierotti, M. *La Musurgia Universalis di Athanasius Kircher. Aspetti della combinatoria nel Seicento: l'Ars magna al servizio dell'arte*. Available from http://www.chierotti.net/kircher/articolo1
31. Galliera, S. (2014). *L'estetica musicale di G.W. Leibniz*. Ph.D. thesis, Università degli Studi di Macerata. Available from http://ecum.unicam/793/

Chapter 24
Fractal Self-similarity: From Geometric Structures to Coherent Dynamics

Giuseppe Vitiello

1 Introduction

In recent works it has been shown that the formalism describing fractal self-similar structures is isomorph to the one of coherent boson condensates, in particular to squeezed SU(1,1) coherent states [1–5]. On the other hand, these last ones provide a representation of a dissipative system made of damped and amplified oscillators.

In space-time regions where the magnetic field may be approximated to be constant and the electric field is derivable from a harmonic potential, the isomorphism is shown to exist also between electrodynamics and the system of damped/amplified oscillators, and thus squeezed SU(1,1) coherent states and fractal self-similar structures [1]. A link is thus established between electrodynamics, dissipation, self-similarity and squeezed coherent states.

The plan of this report is the following. In Sect. 2 the study of the conservation of the energy-momentum tensor in electrodynamics shows that this is isomorph to a set of damped/amplified oscillators. Their quantization and their isomorphism with squeezed coherent states is considered in Sect. 3. In Sect. 4 the isomorphism between fractal self-similarity and squeezed coherent state is established, also showing the isomorphism between these last ones, self-similarity and electrodynamics. Sect. 5 is devoted to conclusions and outlook. I will closely follow [1] in the presentation.

G. Vitiello (✉)
Dipartimento di Fisica "E.R. Caianiello", Università di Salerno and INFN,
Sezione di Napoli, Gruppo Collegato Salerno, Fisciano (Salerno) 84084, Italy
e-mail: vitiello@sa.infn.it

G. Minati et al. (eds.), *Towards a Post-Bertalanffy Systemics*, Contemporary Systems Thinking, DOI 10.1007/978-3-319-24391-7_24

2 The Energy-Momentum Tensor in Electrodynamics

In classical electrodynamics, as well as in quantum electrodynamics, the meaning of the conservation of the electromagnetic (em) energy-momentum tensor $T^{\mu\nu}$, $\partial_\mu T^{\mu\nu} = 0$, is that there is no energy/momentum flowing out of the *closed* system $\{\psi, A^\mu\}$, made of the matter field ψ and of the em gauge field A^μ. However, one may easily realize that the conservation of $T^{\mu\nu}$ arises from the compensation of the variations of the matter part $T_m^{\mu\nu}$ and the em part $T_\gamma^{\mu\nu}$ of the total $T^{\mu\nu}$: $\partial_\mu T_m^{\mu\nu} = -\partial_\mu T_\gamma^{\mu\nu}$, so that $\partial_\mu T^{\mu\nu} = \partial_\mu (T_m^{\mu\nu} + T_\gamma^{\mu\nu}) = 0$. In particular one finds:

$$\partial_\mu T_m^{\mu\nu} = e F^{\alpha\nu} J_\alpha \tag{24.1}$$

$$\partial_\mu T_\gamma^{\mu\nu} = -e F^{\alpha\nu} J_\alpha \tag{24.2}$$

where J_α denotes the current, e is the charge and as usual $F^{\alpha\beta} = \partial^\beta A^\alpha - \partial^\alpha A^\beta$ ($g^{\mu\nu} = (1,-1,-1,-1)$, $\mu = 0,1,2,3$; $\hbar = 1 = c$). We see that the non-vanishing divergences of $T_m^{\mu\nu}$ and $T_\gamma^{\mu\nu}$ compensate each other. For what concerns our discussion there is no need to specify the boson or fermion nature of $\psi(x)$. Volume integration of Eqs. (24.1) and (24.2) gives for $\nu = 0$ the rate of changes in time of the energy of the matter field and em field, $\mathscr{E}_m$ and $\mathscr{E}_\gamma$, respectively:

$$\partial_0 \mathscr{E}_m = e\mathbf{E} \cdot \mathbf{v} = -\partial_0 \mathscr{E}_\gamma \tag{24.3}$$

For $\nu = i = 1,2,3$, integration of Eqs. (24.1) and (24.2) over the volume gives

$$\partial_0 P_m^i = eE^i + e(\mathbf{v} \times \mathbf{B})^i \tag{24.4}$$

$$\partial_0 P_\gamma^i = -eE^i - e(\mathbf{v} \times \mathbf{B})^i \tag{24.5}$$

namely, the Lorentz forces $\mathbf{F}_m$ and $\mathbf{F}_\gamma$, acting on two opposite charges with same velocity $\mathbf{v}$ in the same electric and magnetic fields, $\mathbf{E}$ and $\mathbf{B}$, respectively.

Suppose now that, at least in some space-time region, the magnetic field $\mathbf{B}$ be a constant vector, thus described by the vector potential

$$\mathbf{A} = \frac{1}{2}\mathbf{B} \times \mathbf{r} \tag{24.6}$$

where $\mathbf{r} = (x_1, x_2, x_3)$. It is $\mathbf{B} = \nabla \times \mathbf{A}$, $\nabla \cdot \mathbf{A} = 0$. Choose the reference frame so that $\mathbf{B} = \nabla \times \mathbf{A} = -B\hat{\mathbf{3}}$. Then, $A_3 = 0$ and by using $\varepsilon_{12} = -\varepsilon_{21} = 1$; $\varepsilon_{ii} = 0$,

$$A_i = \frac{B}{2}\varepsilon_{ij} x_j\,, \qquad i,j = 1,2. \tag{24.7}$$

The third component, $i = 3$, of $(\mathbf{v} \times \mathbf{B})$ vanishes. Assume also that $\mathbf{E}$ is given by the gradient of the harmonic potential $\Phi \equiv \frac{k}{2e}({x_1}^2 - {x_2}^2) \equiv \Phi_1 - \Phi_2$, $\mathbf{E} = -\nabla\Phi$; and $E_3 = 0$. We may thus limit our analysis to the $i = 1,2$ components in Eqs. (24.4) and (24.5). Then let $i = 1$ in Eq. (24.4) and put $B \equiv \gamma/e$. Use of Eq. (24.7) gives

$$m\ddot{x}_1 + \gamma\dot{x}_2 + kx_1 = 0 \tag{24.8}$$

where m, γ and k are time independent quantities and the first member of Eq. (24.4) has been put equal to $m\ddot{x}_1$ (and is equal and opposite to the $i = 1$ component of the force in Eq. (24.5)). By considering $i = 2$ in Eq. (24.5) gives

$$m\ddot{x}_2 + \gamma\dot{x}_1 + kx_2 = 0 \tag{24.9}$$

(note that similar conclusion may be reached considering $i = 2$ in Eq. (24.4) and $i = 1$ in Eq. (24.5)). Equations (24.8) and (24.9) can be derived from the Lagrangian

$$L = \frac{1}{2m}(m\dot{x}_1 + e_1A_1)^2 - \frac{1}{2m}(m\dot{x}_2 + e_2A_2)^2 - \frac{e^2}{2m}({A_1}^2 - {A_2}^2) - e\Phi \tag{24.10}$$

The Hamiltonian is [6, 7]

$$H = H_1 - H_2 = \frac{1}{2m}(p_1 - e_1A_1)^2 + e_1\Phi_1 - \frac{1}{2m}(p_2 + e_2A_2)^2 + e_2\Phi_2 \tag{24.11}$$

In the least energy state (where $H = 0$, $H_1 = H_2$) the respective contributions to the energy compensate each other. One of the oscillators may be considered to represent the em field in which the other one is embedded and vice-versa.

In summary, the system of damped/amplified oscillators Eqs. (24.8) and (24.9) provide, under the conditions specified above, a representation of the content of Maxwell equations and the associated conservation laws.

3 The Quantum Field Theory Framework

In this section I show that the damped/amplified oscillators have a quantum representation in terms of squeezed $SU(1,1)$ coherent states. Thus, the isomorphism between such states and electrodynamics is also established. This is also consistent with a recent result [8] by which the electromagnetic quantum field appears to be described by the $su(1,1)$ algebra.

The system of oscillators considered above belongs to the class of deterministic systems à la 't Hooft [9–13], i.e., those systems that remain deterministic even when described by means of Hilbert space techniques. As shown in [9–13], the quantum harmonic oscillator is obtained provided that a constraint on the Hilbert space of the form $J_2|0\rangle = 0$ is imposed (J_2 is a generator of the $SU(1,1)$ group, see below). The Hamiltonian H for the quantum damped/amplified oscillator system, corresponding to the one given in Eq. (24.11), is found to be [14] (see also [15–17])

$$H = H_0 + H_I \tag{24.12}$$

$$H_0 = \hbar\Omega(A^\dagger A - B^\dagger B), \qquad H_I = i\hbar\Gamma(A^\dagger B^\dagger - AB) \tag{24.13}$$

where $\Gamma \equiv \frac{\gamma}{2m}$, $A \equiv \frac{1}{\sqrt{2}}(a+b)$, $B \equiv \frac{1}{\sqrt{2}}(a-b)$, with the annihilation and creation operators

$$a \equiv \left(\frac{1}{2\hbar\Omega}\right)^{\frac{1}{2}} \left(\frac{p_x}{\sqrt{m}} - i\sqrt{m}\Omega x\right); \quad a^\dagger \equiv \left(\frac{1}{2\hbar\Omega}\right)^{\frac{1}{2}} \left(\frac{p_x}{\sqrt{m}} + i\sqrt{m}\Omega x\right) \quad (24.14)$$

$$b \equiv \left(\frac{1}{2\hbar\Omega}\right)^{\frac{1}{2}} \left(\frac{p_y}{\sqrt{m}} - i\sqrt{m}\Omega y\right); \quad b^\dagger \equiv \left(\frac{1}{2\hbar\Omega}\right)^{\frac{1}{2}} \left(\frac{p_y}{\sqrt{m}} + i\sqrt{m}\Omega y\right) \quad (24.15)$$

and $[a,a^\dagger]=1=[b,b^\dagger]$, $\quad [a,b]=0=[a,b^\dagger]$. As customary we have assumed the commutators $[x,p_x]=i\hbar=[y,p_y]$, $[x,y]=0=[p_x,p_y]$.

By defining n_A and n_B the number of A's and B's and $(A\otimes 1)|0\rangle\otimes|0\rangle \equiv A|0\rangle = 0$; $(1\otimes B)|0\rangle\otimes|0\rangle \equiv B|0\rangle = 0$, the vacuum state is $|0\rangle \equiv |n_A=0,n_B=0\rangle = |0\rangle\otimes|0\rangle$. Its time evolution is controlled by H_I and given by $|0(t)\rangle = e^{-it\frac{H}{\hbar}}|0\rangle = e^{-it\frac{H_I}{\hbar}}|0\rangle$, with $\langle 0(t)|0(t)\rangle = 1$, $\forall t$. The evolutor operator and $|0(t)\rangle$ are given by

$$\mathscr{U}(t) = \prod_\kappa \exp\left(\Gamma_\kappa t\left(A_\kappa^\dagger B_\kappa^\dagger - A_\kappa B_\kappa\right)\right) \quad (24.16)$$

$$|0(t)\rangle = \prod_\kappa \frac{1}{\cosh(\Gamma_\kappa t)} \exp\left(\tanh(\Gamma_\kappa t) A_\kappa^\dagger B_\kappa^\dagger\right)|0\rangle \quad (24.17)$$

respectively. We have

$$\lim_{t\to\infty}\langle 0(t)|0\rangle \propto \lim_{t\to\infty}\exp(-t\Gamma) = 0 \quad (24.18)$$

and, in the infinite volume limit, for $\int d^3\kappa\, \Gamma_\kappa$ finite and positive,

$$\langle 0(t)|0\rangle \to 0 \text{ as } V\to\infty \;\forall\, t \quad (24.19)$$

One also has that $\langle 0(t)|0(t')\rangle \to 0$ as $V\to\infty$ $\forall\, t$ and t', $t'\neq t$. This means that a representation $\{|0(t)\rangle\}$ of the canonical commutation relations (CCR) is defined at each time t and is unitarily inequivalent to any other representation $\{|0(t')\rangle, \forall t'\neq t\}$ in the infinite volume limit. The system thus evolves in time through unitarily inequivalent representations of CCR [14]. The number of A_κ (or B_κ) modes in $|0(t)\rangle$ is given by

$$\mathscr{N}_{A_\kappa}(t) = \langle 0(t)|A_\kappa^\dagger A_\kappa|0(t)\rangle = \sinh^2\Gamma_\kappa t \quad (24.20)$$

Define $J_+ \equiv A^\dagger B^\dagger$, $\quad J_- \equiv J_+^\dagger \equiv AB$, $\quad J_3 \equiv \frac{1}{2}(A^\dagger A + B^\dagger B + 1)$, then $[J_+,J_-] = -2J_3$, $\quad [J_3,J_\pm] = \pm J_\pm$, which provides the two mode realization of the algebra $su(1,1)$, with $SU(1,1)$ Casimir operator $\mathscr{C}$ given by $\mathscr{C}^2 = \frac{1}{4}(A^\dagger A - B^\dagger B)^2$, so that $[H_0,H_I]=0$. This last commutation relation guarantees that the initial condition of positiveness for the eigenvalues of H_0 is protected against transitions to negative energy states.

We thus recognize that $|0(t)\rangle$ is a two-mode time dependent generalized $SU(1,1)$ coherent state [14, 15, 18, 19]. In particular, $|0(t)\rangle$ is a squeezed coherent state characterized by the q-deformation of Lie-Hopf algebra [15, 20–22]. Note that A and B are entangled modes. This is consistent with the entanglement between the (charged) matter field and the em field in QED. I remark that the entanglement cannot be destroyed by the action of any unitary operator, a feature absent in quantum mechanics.

Finally, I observe that states generated by $B^\dagger$ represent the sink where the energy dissipated by the quantum damped oscillator flows, so that the B-oscillator represents the reservoir coupled to the A-oscillator [14]. It can be shown that time evolution is controlled by entropy variations [14, 23], which is consistent with the fact that dissipation implies breaking of time-reversal invariance, namely a privileged direction in time evolution, *the arrow of time*. The variations in time of the number of particles condensed in the vacuum gives heat dissipation $dQ = \frac{1}{\beta} dS$.

4 On the Isomorphism Between Fractal Self-similarity and Squeezed Coherent States

In this section it will be shown that an isomorphism exists between the fractal self-similarity properties and the $SU(1,1)$ squeezed coherent states [1–5], which, on the basis of the results of the previous section, also establishes the isomorphism between electrodynamics and fractal self-similarity.

Let me consider the logarithmic spiral (Fig. 24.1). The results can be extended to other examples of deterministic fractals (i.e., those which are generated iteratively according to a prescribed recipe), such as the Koch curve, the Sierpinski gasket and carpet, the Cantor set, etc. [24, 25]. In polar coordinates (r, θ) the logarithmic spiral [25, 26] is represented by:

$$r = r_0 \, e^{d\,\theta} \tag{24.21}$$

r_0 and d and $r_0 > 0$ are arbitrary real constants. In a log-log plot with abscissa $\theta = \ln e^{\theta}$, Eq. (24.21) is represented by the straight line of slope d :

$$d\,\theta = \ln \frac{r}{r_0} \tag{24.22}$$

The constancy of the angular coefficient $\tan^{-1} d$ represents the self-similarity property. Rescaling $\theta \to n\,\theta$ affects r/r_0 by the power $(r/r_0)^n$. Consider the associated parametric equations:

$$\xi = r(\theta)\cos\theta = r_0\, e^{d\,\theta} \cos\theta \tag{24.23}$$

$$\eta = r(\theta)\sin\theta = r_0\, e^{d\,\theta} \sin\theta \tag{24.24}$$

In the complex z-plane, the sign of $d\,\theta$ fully specifies the point $z = \xi + i\eta = r_0\, e^{d\,\theta}\, e^{i\,\theta}$ on the spiral. The points $z_1 = r_0\, e^{-d\,\theta}\, e^{-i\,\theta}$ and $z_2 = r_0\, e^{+d\,\theta}\, e^{+i\,\theta}$ need both to be considered due to the completeness of the (hyperbolic) basis $\{e^{-d\,\theta}, e^{+d\,\theta}\}$. For convenience, opposite signs for the imaginary exponent $\pm i\,\theta$ are also considered. z_1 and z_2 are found to solve the equations

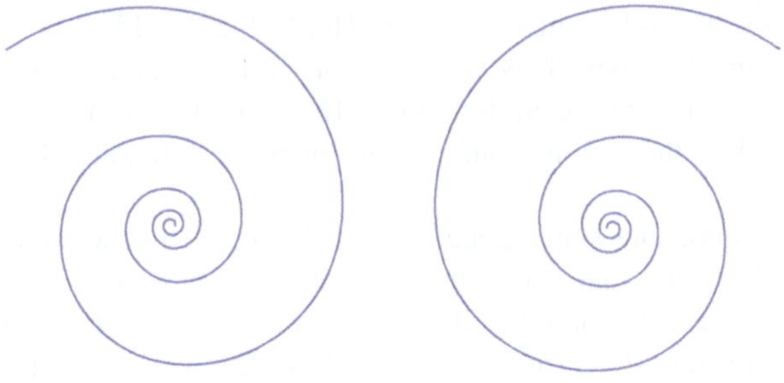

Fig. 24.1: The anti-clockwise and the clockwise logarithmic spiral

$$m\ddot{z}_1 + \gamma \dot{z}_1 + \kappa z_1 = 0 \quad (24.25)$$

$$m\ddot{z}_2 - \gamma \dot{z}_2 + \kappa z_2 = 0 \quad (24.26)$$

respectively, where θ has been parameterized as $\theta(t)$, "dot" denotes derivative with respect to t, and it has been put

$$\theta(t) = \frac{\gamma}{2md}t = \frac{\Gamma}{d}t \quad (24.27)$$

up to an arbitrary additive constant; m, γ and κ are positive real constants. Thus, $z_1(t) = r_0\, e^{-i\Omega t}\, e^{-\Gamma t}$ and $z_2(t) = r_0\, e^{+i\Omega t}\, e^{+\Gamma t}$ solutions of Eqs. (24.25) and (24.26) describe the logarithmic spirals and the parameter t can be interpreted as the time parameter. As in Sect. 3, the notations are $\Gamma \equiv \gamma/2m$ and $\Omega^2 = (1/m)(\kappa - \gamma^2/4m) = \Gamma^2/d^2$, with $\kappa > \gamma^2/4m$. The spiral "angular velocity" is $|d\theta/dt| = |\Gamma/d|$.

Observe that the discrete group of transformations $z_1(m) = r_0(e^{-2\pi d})^m \to z_1(m+1) = r_0(e^{-2\pi d})^{(m+1)} = z_1(m)(e^{-2\pi d})$, due to the T integer multiplicity, $\theta(T) = 2\pi$ at $T = 2\pi d/\Gamma$ and, at $t = mT$, $z_1 = r_0\,(e^{-2\pi d})^m$, $z_2 = r_0\,(e^{2\pi d})^m$, with integer $m = 1, 2, 3 \ldots$, is included in the continuous time evolution (cf. Eq. (24.27)). The isomorphism seems thus to appears as a homomorphism.

Our discussion also includes the golden spiral and its relation with Fibonacci progression, which here for brevity are not considered.

Note that by putting $[z_1(t) + z_2^*(-t)]/2 = x(t)$ and $[z_1^*(-t) + z_2(t)]/2 = y(t)$, Eqs. (24.25) and (24.26) reduce to Eqs. (24.8) and (24.9), namely they provide an equivalent representation of Eqs. (24.4) and (24.5). Thus, the Lagrangian, the Hamiltonian, the evolution operator, the ground state, etc. are obtained for the spiral when working in the proper QFT frame in a similar way as done in previous sections [1]. The $SU(1,1)$ generalized coherent state (24.17) is a squeezed state and again the choice of a privileged direction in time evolution emerges from the breakdown of time-reversal symmetry and one can define the entropy operator S. The time-reversed, *but distinct*, image of the right-handed chirality spiral (indirect spiral, $q < 1$) is the left-handed chirality spiral (direct spiral, $q \equiv e^{-d\theta} > 1$). Without going in the details, I also mention that the Hamiltonian H is actually the *fractal free energy* for the process of coherent boson condensation out of which the fractal is formed and the system temperature $T = \hbar\Gamma$ is found to be proportional to the background zero point energy: $\hbar\Gamma \propto \hbar\Omega/2$ [11–13, 15].

These results can be extended to the Koch curve and to other fractals [2]. For brevity their derivation is not reported here. Summarizing, the dynamical description of the geometrical feature of fractal self-similarity is obtained.

By using the notation $+ \equiv 1$ and $- \equiv 2$, and letting $p_{z\pm}$ denote the momenta and $v_\pm = \dot{z}_\pm$ the forward in time and backward in time velocities, we have:

$$v_\pm = \pm\frac{1}{m}\left(p_{z\pm} \mp (1/2)\gamma z_\mp\right), \qquad \text{with} \qquad [v_+, v_-] = i\frac{\gamma}{m^2} \tag{24.28}$$

The canonical set of conjugate position coordinates (ξ_+, ξ_-) can be defined by putting $\xi_\pm = \mp(m/\gamma)v_\mp$, with

$$[\xi_+, \xi_-] = i\frac{1}{\gamma} \tag{24.29}$$

which provides a hint on how to proceed in order to show the relation between dissipation and noncommutative geometry in the plane (z_1, z_2) and the role of the deformed Hopf algebra [1, 27]. However, for brevity I will not discuss further such an issue.

5 Towards an Integrated Vision of Nature

A link is established between electrodynamics, self-similarity and coherent states in QFT. In space-time regions where the magnetic field may be approximated to be constant and the electric field is derivable from a harmonic potential, an isomorphism has been recognized to exist between electrodynamics and a set of damped and amplified oscillators, a prototype of a dissipative system and its environment, respectively. On the other hand, such a system of oscillators is isomorph to fractal self-similar structures [1–5] and is represented by $SU(1,1)$ squeezed (q-deformed) coherent states. Dissipation plays a central role and is at the origin of noncommutative geometry in the plane and is described by the deformed Hopf algebra [3–5, 20–22, 28, 29].

Fractal-like structures with self-similarity properties appear as *macroscopic quantum systems* generated by coherent $SU(1,1)$ quantum condensation processes at the microscopic level [15, 30–32]. Fractals emerge out of a process of *morphogenesis* (*forms*) as the macroscopic manifestation of the dissipative, coherent quantum dynamics at the elementary level. An integrated vision of Nature based on the dynamics of coherence thus emerges. It includes also the sector of high energy physics, with the coherent condensate structure of the vacuum and the recent discovery of the Higgs boson belongs to such a picture. Nature appears to be shaped by coherence, rather than being organized in isolated compartments, in collections of isolated systems [2, 33]. The dynamics of coherence appears to be the primordial origin of *codes*. These are lifted from the (syntactic) level of pure information (à la Shannon) to the (semantic) level of *meanings*, expressions of coherent dynamical processes. Codes, (including the genetic DNA code [1, 34]) appear to be the *vehicles* through which coherence propagate and manifest itself.

References

1. Vitiello, G. (2014). On the isomorphism between dissipative systems, fractal self-similarity and electrodynamics. Toward an integrated vision of nature. *Systems, 2*, 203–216. doi:10.3390/systems2020203.
2. Vitiello, G. (2012). Fractals, coherent states and self-similarity induced noncommutative geometry. *Physics Letters, A376*, 2527–2532.
3. Vitiello, G. (2009). Coherent states, fractals and brain waves. *New Mathematics and Natural Computation, 5*, 245–264.
4. Vitiello, G. (2009). Fractals and the Fock-Bargmann representation of coherent states. In W. Lawless, K. van Rijsbergen, & M. Klusch (Eds.), *Quantum interaction. Lecture notes in artificial intelligence* (pp. 6–16). Berlin/Heidelberg: Springer.
5. Vitiello, G. (2011). Topological defects, fractals and the structure of quantum field theory. In I. Licata & A.J. Sakaji (Eds.), *Vision of oneness* (pp. 155–179). Roma, Italy: Aracne Edizioni.
6. Celeghini, E., Graziano, E., Nakamura, K., & Vitiello, G. (1992). Finite temperature quantum field theory and gauge field. *Physics Letters, B285*, 98–102.
7. Blasone, M., Graziano, E., Pashaev, O.K., & Vitiello, G. (1996). Dissipation and topologically massive gauge theories in the pseudo-Euclidean plane. *Annals of Physics, 252*, 115–132.
8. Marzuoli, A., Raffa, F.A., & Rasetti, M. (2014). Where do bosons actually belong? *Journal of Physics A Mathematical and Theoretical, 06*, 47(27). doi:10.1088/1751-8113/47/27/275202.
9. 't Hooft, G. (1999). Quantum gravity as a dissipative deterministic system. *Classical and Quantum Gravity, 16*, 3263–3279.
10. 't Hooft, G. (2007). A mathematical theory for deterministic quantum mechanics. *Journal of Physics Conference Series, 67*, 012015. doi:10.1088/1742-6596/67/1/012015.
11. Blasone, M., Jizba, P., & Vitiello, G. (2001). Dissipation and quantization. *Physics Letters, A287*, 205–210.
12. Blasone, M., Celeghini, E., Jizba, P., & Vitiello, G. (2003). Quantization, group contraction and zero point energy. *Physics Letters, A310*, 393–399.
13. Blasone, M., Jizba, P., Scardigli, F., & Vitiello, G. (2009). Dissipation and quantization in composite systems. *Physics Letters, A373*, 4106–4112.
14. Celeghini, E., Rasetti, M., & Vitiello, G. (1992). Quantum dissipation. *Annals of Physics, 215*, 156–170.
15. Blasone, M., Jizba, P., & Vitiello, G. (2011). *Quantum field theory and its macroscopic manifestations* London, UK: Imperial College Press.
16. Blasone, M., Srivastava, Y. N., Vitiello, G., & Widom, A. (1998). Phase coherence in quantum Brownian motion. *Annals of Physics, 267*, 61–74.
17. Srivastava, Y. N., Vitiello, G., & Widom, A. (1995). Quantum dissipation and quantum noise. *Annals of Physics, 238*, 200–207.
18. Perelomov, A. (1986). *Generalized coherent states and their applications*. Berlin: Springer.
19. Klauder, J. R., & Skagerstam, B. (1985). *Coherent states*. Singapore: World Scientific.
20. Celeghini, E., De Martino, S., De Siena, S., Rasetti, M., & Vitiello, G. (1995). Quantum groups, coherent states, squeezing and lattice quantum mechanics. *Annals of Physics, 241*, 50–67 .
21. Celeghini, E., Rasetti, M., Tarlini, M., & Vitiello, G. (1989). $SU(1,1)$ squeezed states as damped oscillators. *Modern Physics Letters, B3*, 1213–1220.

22. Celeghini, E., Rasetti, M., & Vitiello, G. (1991). On squeezing and quantum groups. *Physical Review Letter, 66*, 2056–2059.
23. De Filippo, S., & Vitiello, G. (1977). Vacuum structure for unstable particles. *Letter Nuovo Cimento, 19*, 92–96.
24. Bunde, A., & Havlin, S. (1995). *Fractals in science* Berlin: Springer
25. Peitgen, H. O., Jürgens, H., & Saupe, D. (1986). *Chaos and fractals: New frontiers of science* Berlin: Springer.
26. Andronov, A. A., Vitt, A. A., & Khaikin, S. E. (1966). *Theory of oscillators* Mineola: Dover Publication.
27. Sivasubramanian, S., Srivastava, Y. N., Vitiello, G., & Widom, A. (2003). Quantum dissipation induced noncommutative geometry. *Physics Letter, A311*, 97–105.
28. Celeghini, E., De Martino, S., De Siena, S., Iorio, A., Rasetti, M., & Vitiello, G. (1998). Thermo field dynamics and quantum algebras. *Physics Letters, A244*, 455–461.
29. Vitiello, G. (2007). Links. Relating different physical systems through the common QFT algebraic structure. In W. G. Unruh & R. Schuetzhold (Eds.), *Quantum analogues: From phase transitions to black holes and cosmology. Lectures notes in physics* (Vol. 718, pp. 165–205). Berlin: Springer.
30. Umezawa, H., Matsumoto, H., & Tachiki, M. (1982). *Thermo field dynamics and condensed states*. Amsterdam: North-Holland.
31. Umezawa, H. (1993). *Advanced field theory: Micro, macro and thermal concepts*. Melville: American Institute of Physics.
32. Bunkov, Y. M., & Godfrin, H. (2000). *Topological defects and the nonequilibrium dynamics of symmetry breaking phase transitions*. Dordrecht: Kluwer Academic Publishers.
33. Vitiello, G. (2012). Struttura e Funzione. Una visione ecologica integrata. *Rivista di Filosofia Neo-Scolastica, 4*, 625–637.
34. Montagnier, L., Aissa, J., Del Giudice, E., Lavallee, C., Tedeschi, A., & Vitiello, G. (2011). DNA waves and water. *Journal of Physics: Conference Series, 306*, 012007. doi:10.1088/1742-6596/306/1/012007.

Chapter 25
Enhancement in Mathematical Abilities: A System Approach

Maria Lidia Mascia, Mirian Agus, Maria Chiara Fastame, and Alessandra Addis

1 Introduction

Mathematical learning is always included in educational programs. It involves a lot of factors that are fundamental to have success in this subject, however, a large number of studies highlights the difficulties that are often correlated to the mathematical learning. A body of literature evidences how children have their very first mathematical experience through the vision and the manipulation of concrete objects [16, 17, 19]. The numerical intelligence is the ability to manipulate numerical quantities, recognizing and distinguishing them. The number is specifically a cognitive domain that requires intelligence and competence [17]. There are several theories that explain the ways in which you get to learn the number and quantity. Some studies assert that the concept of number is innate, for example this point of view is supported by Butterworth [8, 9], while other authors, such as Piaget (1958 in Siegler [31]) affirm the opposite. The most recent researches have widely demonstrated that infants are capable of discriminating different sets of numbers, so it is important to understand the role of the environment and the system in the promotion and development of that potential. The present study has acted on a number of variables. These variables had the aim to promote the best development of emerging skills and to prevent some difficulties related to the development of mathematical abilities.

M.L. Mascia (✉) · M. Agus · M.C. Fastame · A. Addis
Department of Pedagogy, Psychology, Philosophy - Faculty of Humanistic Studies, University of Cagliari, Via Is Mirrionis 1, 09123 Cagliari, Italy
e-mail: marialidia.mascia@unica.it

G. Minati et al. (eds.), *Towards a Post-Bertalanffy Systemics*, Contemporary Systems Thinking, DOI 10.1007/978-3-319-24391-7_25

2 The Role of Enhancement in Cognition and Metacognition

The main role in the enhancement in mathematical abilities is represented by the interventions which support the development of a function in its emergence [27]. Recent numerical studies of intelligence state that the ability to distinguish the number, to understand and operate on the quantitative aspects of reality is an innate potential in children [17]. These development processes don't have to be only spontaneous, but sometimes they require educational strategies designed to strengthen them and to encourage their development. The numeric intelligence or the ability to manipulate the knowledge through the complex cognitive system of numbers [17], it isn't a defined entity, but a potential that can be developed and made efficient through paths of enhancement. These paths have to act on infant's basic cognitive skills and they have to teach him the use of strategies in order to maximize his resources and his supporting components, emotional and motivational states [10]. As Hildt says cognitive enhancement as the attempt to increase cognitive functions has received considerable attention during the last decade [14]. The enhancement is the product of the combination of cognitive and metacognitive factors. Among cognitive factors, the visuo-spatial component has a very important role in the development of mathematical abilities. In recent decades, research has highlighted the importance of visual-spatial skills and visual-spatial working memory for school learning and, in particular, in mathematics, the relationship between space and number is fundamental. Piaget's theory stated that space was an obstacle in the understanding of number concepts because the spatial properties of the objects appeared to interfere with the appreciation of the numerical properties; this point of view has been refuted by many experimental studies, on the assumption of the mental number line [6]. On the other hand a crucial aspect in the enhancement of mathematical abilities is played by a metacognitive component, for instance, in the numerical task it can facilitate the choice of the strategy to use [30]. The metacognitive component includes the knowledge of the person, the knowledge of nature of the task and the characteristics of the strategies that can be implemented to enhance cognitive behavior, goals, tasks and actions necessary to achieve stated aims [5]. In recent years, many studies centre on the role of metacognition in mathematics education [7]. Butler and Winn, for example, [7] describe how evaluating goals, thinking of strategies and choosing is the most appropriate strategy for solving the problem. In this model, feedback is a central factor in cognitive–metacognitive processes [15]. The metacognitive aspect is related to the importance of control procedure such as monitoring the comprehension, controlling the execution plan, evaluating the results in a mathematical assignment [20]. A new trend of research has suggested some educational solutions aimed at favouring the scholastic achievements of children with typical and atypical developmental trajectories. For this reason several studies suggest the use of trainings to enhance specific cognitive processes in childhood.

3 The Role of Technological Tools

The use of technological tools is important to create a system approach in the development and enhancement in mathematical abilities. The introduction of technologies as support of educational activities shows how fundamental are computer-assisted interventions in the field. In an effort to engage children in mathematics learning, many primary teachers use mathematical games and activities [15]. The use of technology is one of intervention that can be implemented to design a comprehensive educational program. In literature, the use of educational software and multimedia is effective to support the work both of the student and the educator/teacher. It involves many advantages. Multi-media interests different cognitive channels through the simultaneous presentation of some stimuli, for example, images, text, animations, sounds and voices. So, electronic technologies are essential tools for teaching, learning and doing mathematics. When technological tools are available, students can focus on decision making, reflection, reasoning and problem solving [15]. In our study we use the educational software. The educational software has a huge range of strengths [1]. Among them, a fundamental element is the possibility of combining elements perceptual, attentional and visual-spatial memory. The educational software has also the advantage of customizing the paths and present the elements in an accessible and usable way. Another particular aspect is the child's motivation, the characters guide (avatar) is able to approach the child to the task. In addition, the use of computers allows a more standardized presentation of the stimuli and an accurate record of the responses. In addition, the program allows an easy handling and printing of stored data. The guidelines for the implementation of the software will be mainly: ease of use, speed and efficiency, agreeableness, the average technology and the interchangeability [2]. One of the more complete theories of learning with multimedia is the theory of Mayer [21]. According to the author we can approach the study of learning media following two approaches. One is based on the technology in which the main role is played by the software and learning is seen as a passive process; the other approach focuses primarily on the individual, studying basic cognitive processes and the way in which it interacts with the system, in this case, learning is seen as an active process [12].

4 A System Approach

The importance of understanding the development of mathematical cognition has long been recognized, both for the appreciation of the emergence of numerical skills and what it can tell us more generally about cognitive development [22]. However, a large group of studies have separately explored the role that some cognitive components have on the learning of mathematics (working memory, attention, motivation, etc ...) [23]. The literature presents a small number of studies that relates to cognitive processes and metacognitive learning strategies, the presentation of mathematical learning programs (paper and pencil and/or educational software)

[32] and background variables that are the foundation of mathematics learning [4]. To this end, a process has been studied, with the help of researchers, students, parents, teachers, tools and methodologies in order to endorse step by step some key elements of the computing ability, fundamental to support the "phase transition" [24, 25]. Furthermore, a large body of literature suggests that complete psycho-educational programs for the empowerment of cognitive and metacognitive functions can contribute to the school achievements, both for typically developing children and those showing learning disabilities.

5 The Research

The aim of the present study was to evaluate the impact of single and combined mathematical trainings, respectively presented in computer-assisted or pencil-and-paper modalities on the development of numeracy skills in children of 7–8 years old attending the third primary school of some comprehensive schools of Cagliari. This work would also like to highlight the importance of the relationship between the numerical intelligence and visual-spatial memory.

5.1 Subjects

One hundred and forty-four children of 7–8 years old are recruited in several primary schools located in Sardinia (Italy). The participants (82 male and 62 female) were divided into six experimental groups (and a control group) that followed one or combined type of trainings, in pencil-and-paper and/or computerized formats.

5.2 Method

Participants had followed trainings for enhancing numerical abilities (paper and/or multimedia) created by Lucangeli et al. [18] and/or visuo-spatial abilities by Fastame and Antonini [13] to test the effectiveness of the enhancements, visual-spatial skills (using the CPM [3]) and calculation (using the ACMT 6–11 [11]) were detected before and after the trainings. specifically, the experimental groups carried out the following treatments: the first experimental group (G1) carried out the mathematical training in computerised format and visuo-spatial training in computerized format ($n = 14$); G2 was presented with the mathematical training in pencil-and-paper format ($n = 8$); G3 was given the visuo-spatial training in the pencil-and-paper format ($n = 25$); G4 mathematical training in computerized format and visuo-spatial training in the pencil-and-paper format ($n = 18$); G5 carried out the mathematical both in the pencil-and-paper and computerized formats ($n = 18$); G6 was presented the visuo spatial training in computerized modality and the mathematical training in the pencil-and-paper format ($n = 19$). The control group was made up of 25 pupils. The teachers and their pupils voluntarily participated in the research (non-probabilistic sampling).

5.3 Results

We analyzed the data from the six groups. It was computed an index of gain in order to highlight the changes occurred in each dimension between the detection pre-test and post-test (value obtained by the post-test–value obtained pre)/value obtained at the pre-test. On such data was applied MANCOVA, having as a factor the activity achieved. The results show the effect of the training factor in relation to the dimension of numerical knowledge. We calculated the gain between post and pre evaluations, in relation to each scale of AC-MT (*Written calculation, Speed, Accuracy, Semantic and Syntactic numerical knowledge*). The multivariate tests were significant for the covariates and for the factor "training group". Then univariate tests indicated a significant effect of "training group" in terms of *Semantic and Syntactic numerical knowledge* ($F_{(6,133)} = 3{,}766$, $p = 0.042$).

6 Discussion

The main goal of the present research was to investigate the part played by single or combined mathematical trainings in favoring the enrichment of numerical knowledge. Therefore, computerised-assisted and/or paper-and-pencil single or combined treatments were proposed to 144 children attending the third of some primary schools in Cagliari.

The results of the research showed that if properly stimulated (with different types of learning tools) the resources of the subject may support an improvement in their skills.

The use of computer-assisted software is very effective in empowering cognitive abilities in primary school, our study suggests that the use of the computer is not sufficient to enrich numeracy knowledge, the results shows the efficacy of a set of combined elements. Indeed, we found that children trained by a computer-assisted and pencil-and-paper mathematical treatment outperformed children trained by a combined computer-based and pencil-and-paper visuo-spatial trainings in terms of processing speed for number manipulation. The modality (i.e., computer-assisted versus pencil-and-paper) in which the trainings were presented, the combination of visuo-spatial and mathematical materials was as much effective as the "pure" combined (i.e., computer-assisted and pencil-and-paper) mathematical treatment. So, the mathematical training is particularly useful when combined with the visual-spatial [28]. In order to investigate such issues more closely, it would be useful to extend the number of children and to compare other different combinations of training. We can conclude that our results confirm the findings of previous researches [26, 29].

Our results shows that when properly stimulated, the subject's resources can support the improvement of individual skills [17]. The teacher should be aware of all the instruments and of all cognitive and metacognitive processes. These tools are essential for the school and families. The design and implementation of educational interventions with a systemic vision aimed at producing a cognitive enhancement that can be effective in the course of time.

References

1. Alessi, S. M., & Trollip, S. R. (2001). Multimedia for learning. *Methods and Development*, (3rd edition). Boston: Allyn & Bacon.
2. Antonietti, A., & Cantoia, M. (2001). *Imparare con il computer. Come costruire contesti di apprendimento per il software*. Trento: Erickson.
3. Bellacchi, C., Scalisi, T.G., Cannoni, E., & Cornoldi, C. (2008). *CPM - Coloured progressive matrice*. Firenze: OS.
4. Boekaerts, M., & Cascallar, E. (2006). How far have we moved toward the integration of theory and practice in self-regulation? *Educational Psychology Review, 18*(3), 199–210.
5. Borkowski, J. G. (1996). Metacognition: Theory or chapter heading? *Learning and Individual Differences, 8*(4), 391–402.
6. Briyant, P., & Squire, S. (2001). Children's mathematics: Lost and found in space. In M. Gattis (Ed.), *Spatial schemas and abstract thought*. Cambridge: MIT Press.
7. Butler, D. L., & Winne, P. H. (1995). Feedback and self-regulated learning: A theoretical synthesis. *Review of Educational Research, 65*(3), 245–281.
8. Butterwort, B. (1999). *The mathematical brain*. London: Macmillan (Italian Translation, *Intelligenza Matematica*. Milano: Rizzoli).
9. Butterworth, B. (2005). The development of arithmetical abilities. *Journal of Child Psychology and Psychiatry, 46*, 3–18.
10. Cornoldi, C. (2007). *Difficoltà e disturbi dell'apprendimento*. Bologna: Il Mulino.
11. Cornoldi, C., Lucangeli, D., & Bellina, M. (2012). *Test AC-MT 6-11*. Trento: Erickson.
12. De Beni, R., Pazzaglia, F., Molin, A., & Zamperlin, C. (2003). *Psicologia cognitiva dell'apprendimento. Aspetti teorici e applicazioni*. Trento: Erickson.
13. Fastame, M. C., & Antonini, R. (2011). *Recupero in … Abilità visuo-spaziali. Percorsi e attività per la scuola primaria e secondaria di primo grado*. Trento: Erickson.
14. Hildt, E. (2013). Cognitive enhancement—A critical look at the recent debate. In *Cognitive enhancement* (pp. 1–14). Netherlands: Springer.
15. Kramarski, B., & Zeichner, O. (2001). Using technology to enhance mathematical reasoning: Effects of feedback and self-regulation learning. *Educational Media International, 38*(2–3), 77–82.
16. Lucangeli, D., Ianniti, A., & Vettore, M. (2007). *Lo sviluppo dell'intelligenza numerica*. Roma: Carocci.
17. Lucangeli, D., & Mammarella, I. C. (2010). *Psicologia della cognizione numerica. Approcci teorici, valutazione e intervento*. Milano: Franco Angeli.
18. Lucangeli, D., Poli, S., & Molin, A. (2011). *Sviluppare l'intelligenza numerica II*. Trento: Erickson.
19. Lucangeli, D., & Tressoldi, P. E. (2002). Lo sviluppo della conoscenza numerica: alle origine del capire i numeri. *Giornale Italiano di Psicologia, 4*, 701–723.
20. Lucangeli, D., Tressoldi, P. E., & Cendron, M. (1998). *SPM-Test delle abilità di soluzione dei problemi matematici*. Trento: Erickson.

21. Mayer, R. E. (2003). The promise of multimedia learning: Using the same instructional design methods across different media. *Learning and Instruction, 13*(2), 125–139.
22. Muldoon, K., Towse, J., Simms, V., Perra, O., & Menzies, V. (2013). A longitudinal analysis of estimation, counting skills, and mathematical ability across the first school year. *Developmental Psychology, 49*(2), 250.
23. Musso, M., Kyndt, E., Cascallar, E., & Dochy, F. (2012). Predicting mathematical performance: The effect of cognitive processes and self-regulation factors. *Education Research International, 2012*, Article ID 250719, 13 pages, 2012. Doi 10.1155/2012/250719.
24. Penna, M. P. (2002). The emergence of strategies in cognitive development. In *Emergence in complex, cognitive, social, and biological systems* (pp. 55–62). New York: Springer.
25. Penna, M. P., & Pessa, E. (1996). *Le interfacce uomo-macchina*. Roma: Di Renzo.
26. Penna, M.P., Stara, V., & Bonfiglio, N. (2002). A systemic proposal on the use of a new technology as a learning tool in school context. In *Emergence in complex, cognitive, social, and biological systems* (pp. 153–157). New York: Springer .
27. Pessa, E. (2002). What is emergence? In *Emergence in complex, cognitive, social, and biological systems* (pp. 379–382). New York: Springer.
28. Praet, M., & Desoete, A. (2014). Enhancing young children's arithmetic skills through non-intensive, computerised kindergarten interventions: A randomised controlled study. *Teaching and Teacher Education, 39*, 56–65.
29. Räsänen, P., Salminen, J., Wilson, A.J., Aunio, P., & Dehaene, S. (2009). Computer-assisted intervention for children with low numeracy skills. *Cognitive Development, 24*(4), 450–472.
30. Siegler, R. S. (1996). *Emerging minds*. New York: Oxford University Press.
31. Siegler, R. S., Liebert, D. E., & Liebert, R. M. (1973). Inhelder and Piaget's pendulum problem: Teaching preadolescents to act as scientists. *Developmental Psychology, 9*(1), 97.
32. Stillman, G. (2014). Metacognition. In *Encyclopedia of mathematics education* (pp. 445–447). Dordrecht: Springer.

Chapter 26
The Effect of Written Approval on Pupils' Academic and Social Behavior: An Exploratory Study in a Northern Italian Middle School

Dolores Rollo, Francesco Sulla, Mia A. Massarini, and Silvia Perini

1 Introduction

A teacher and his/her classroom in interaction is a complex system: it is impossible to consider one aspect of their relationship in isolation from the rest. A considerable amount of research and demonstrational studies, carried out over the past 50 years, has consistently shown that teacher behavior may be a powerful influence on the behavior of both individual students and whole classes [16]. It has been clearly and unequivocally demonstrated, in a variety of educational contexts and settings, that such key teacher behaviors as contingent praise/approval and reprimand/disapproval may be systematically deployed by teachers so as to increase both academic and appropriate social behaviors and to decrease inappropriate behaviors (e.g. [2, 11, 13]).

Regarding teacher pupils interaction, Schwieso and Hastings [15] acknowledge that

> it is a little obvious to say that teaching is an interactive process

but observe that it is a point often ignored in the research into the complexities of the classroom (p. 124). In their discussion of teachers' rates of approval and disapproval in the classroom they emphasize the importance of the relationship between teacher behavior and student behavior.

D. Rollo (✉)
Dipartimento di Neuroscienze, Università degli Studi di Parma, Parma, Italy
e-mail: dolores.rollo@unipr.it

F. Sulla · S. Perini
Dipartimento di Lettere, Arti, Storia e Società, Università degli Studi di Parma, Parma, Italy
e-mail: francesco.sulla@nemo.unipr.it; silvia.perini@unipr.it

M.A. Massarini
Istituto Comprensivo di Castel San Giovanni e Sarmato "Cardinale Agostino Casaroli", Castel San Giovanni (PC), Italy
e-mail: miaastrid.massarini@studenti.unipr.it

G. Minati et al. (eds.), *Towards a Post-Bertalanffy Systemics*, Contemporary Systems Thinking, DOI 10.1007/978-3-319-24391-7_26

> Teachers' approvals and disapprovals may have some effect on pupils, but they are themselves in part the effects or consequences of pupils' actions: teachers do not approve or disapprove in vacuo (p. 124).

Or as Brophy [4] puts it,

> ... much teacher praise is reactive to and under the control of student behavior rather than vice versa (p. 5).

A more balanced perspective, perhaps, may be that of Nafpaktitis et al. [14] who state that in the feedback system of the classroom,

> students continually influence teacher behavior and vice versa (p. 366).

So far, behavioral work has had its focus on verbal approval and disapproval, although some investigators (e.g. [3, 6]) have included the measurement of non-verbal behaviors. These are usually defined as facial expressions, head nods, etc. Yet when much of the teachers' time is spent on marking pupils' work, it is surprising not to find research on the effects of teachers' written comments, ticks, etc. There is surely considerable scope here for influencing pupils' behavior, the nearest approach having been the use of "a letter home saying how well the pupil has done", as reported by Harrop and McCann [9, 10]. The marks and the writing that teachers put on pupils' books can be interpreted as conveying approval or disapproval certainly as easily as can teachers' comments, and since they are not transient, the likelihood is that they can be more accurately interpreted than comments. Moreover, they have the potential for being witnessed by a different population—i.e. the parents rather than the other pupils in the classroom.

Increasing teacher verbal approval has been shown to produce both increased pupil "on-task" behavior (e.g. [11]) and academic achievement (e.g. [19]). However, as we have said, not much is known about the effects of written approval.

The aim of this exploratory study was to investigate the effect of written approval on pupils' academic performance and on-task behavior.

2 Method

The participants were two teachers and their classrooms (2 year 7 classes, respectively made up of 21 and 23 pupils, and 2 year 9 classes, respectively made up of 21 and 19 pupils) from a northern Italian middle school.

A multiple baseline design across participants was employed in order to guarantee that each child would receive written approval (independent variable) by their teacher for his/her performance in five standardized tests on Italian grammar. Children in the experimental group received extra positive comments/praise for doing a specific part well irrespective of the numerical grade. The dependent variable for this study was the number of correct answers and the percentage of time on-task.

Following a baseline phase in which we wanted to measure pupils' performance before any intervention, the introduction of written approval was staggered across little groups. After the teachers had scored the first test, each class was split in three groups similar for mean and standard deviation. The first group received written approval from the second test onwards; the second group received written approval from the third test onwards; the third group received written approval from the fourth test onwards. In order to guarantee the blindness of the procedure, the teachers were told by the project supervisor which tests had to be marked with extra positive comments after they had scored all of them. The number of correct answers and percentage of time on-task was measured before and after the introduction of written approval in order to see whether any change occurred. Pupils were considered on-task when engaged in behaviors that led to completing the assignment. Observations were made using an adapted version of The Pupil Behavior Schedule [12]. The schedule uses a momentary time sampling method. Pupils were observed at 10-min intervals and judged to be either "on-task" or "off-task". Momentary time sampling has been demonstrated to suit collection of time on-task data (e.g. [7, 8]) and to be practical when teachers are required to simultaneously teach and collect data [1]. Before the intervention the teachers had been trained in the use of the paper pencil tool by the project supervisors using videotapes. They practiced recording the data until they reached 90 % reliability. Once reliability was established, teachers scored the lessons, but were aware that two out of five, taken at random, would also be scored independently by the supervisor. Following this the inter-observer agreement was carried out. The subsequent checks produced level of inter-observer agreement at or above 80 %.

3 Results

Data has been analyzed using a general linear mixed model for repeated measures. To capture the variation between subjects both in performance and on-task behavior we included in the model a random effect associated with the intercept for each pupil. Intra class Correlation Coefficients of 0.83 for test results and 0.59 for on-task behavior justified the choice of the model. Indeed a random effect associated with the intercept, in both cases, explained more than 50 % of variance in our unit of analysis (subjects). This does mean that academic (as a number of correct answers) and social performance (as a percentage of on-task behavior) is highly variable between different individuals.

The interaction between group and test results was not significant (Fig. 26.1). Therefore, there was not enough evidence to reject the null hypothesis of equality of population means. There were no differences between younger pupils and older pupils; there were no differences between male and female.

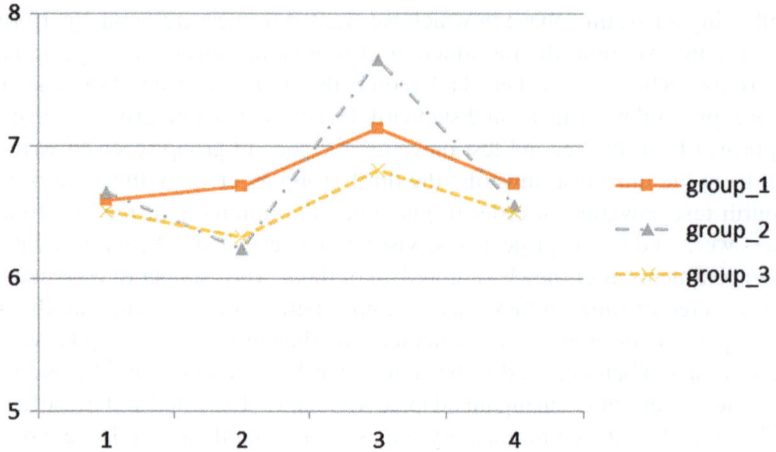

Fig. 26.1: Pupils' number of correct answers in the written test

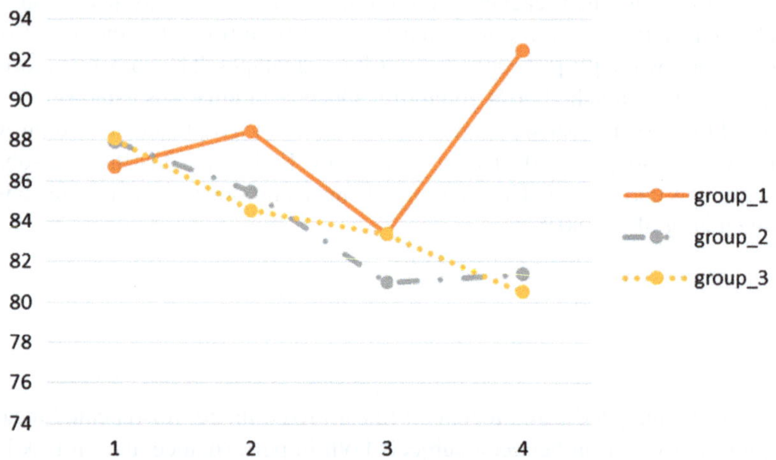

Fig. 26.2: Percentages of pupils' time on-task

Regarding pupils' time on-task, the contrast between the first group (pupils who received written approval from the second test onwards) ($M = 92.50$, $S.D. = 17.59$) and the other two groups (who received the written approval in a staggered fashion: respectively the second group after the third test and the third group after the fourth test) ($M = 81.40$, $S.D. = 23.73$; $M = 80.54$, $S.D. = 27.74$) at the last assessment was significant ($t[56,742] = 2.180$, $p \leq 0.01$) (Fig. 26.2).

4 Discussion

Increasing teacher verbal approval has been shown to produce both increased pupil "on-task" behavior (e.g. [11]) and academic achievement (e.g. [19]). Although much of the teacher's time is spent on marking pupils' work, not too much is known about the effect of written approval on pupils' academic and social behavior. Therefore, the aim of this exploratory study was to investigate the effect of written approval on pupils' academic performance and on-task behavior.

We have not found any effect of written approval on pupils' academic performance measured as a number of correct answers in a standardized Italian grammar test. Among possible explanations for the increased approval group not scoring higher than the other groups might be a possible overlap between approval and numerical grade functions. For some students, only the numerical grade is of interest to them—simple, unambiguous and meaningful in terms of achievement and progression [5]. Taras' [18] suggestion of withholding the grade until students have read and digested the qualitative feedback may reduce that behavior.

Regarding pupils' social behavior measured as a percentage of time on-task, pupils who had received written approval from the second test onwards have shown a growing trend. Although the trend was not significant, on final assessment the first group scored significantly higher than the other groups. These results confirm what Apter et al. [2] have found, studying the effect of verbal approval on pupils' time on-task: approval for academic behavior is one of the most relevant variables which influences pupils' social behavior.

Further study should examine the effect of written approval withholding the numerical grade in order to avoid any interferences. In addition, it would be valuable to identify the conditions necessary for long-term maintenance of the effects of written approval on pupils' time on-task. For example, it is possible that high levels of on-task behavior would be maintained with intermittent teacher written approval.

References

1. Alberto, P. A., & Troutman, A. C. (2012). *Applied behavior analysis for teachers*. New Jersey: Pearson Higher Ed.
2. Apter, B., Arnold, C., & Swinson, J. (2010). A mass observation study of student and teacher behaviour in British primary classrooms. *Educational Psychology in Practice, 26*(2), 151–171.
3. Bain, A., Houghton, S., & Williams, S. (1991). The effects of a school-wide behaviour management programme on teachers' use of encouragement in the classroom. *Educational Studies, 17*(3), 249–260.
4. Brophy, J. (1986). Teacher influences on student achievement. *American Psychologist, 41*(10), 1069.
5. Ecclestone, K. (1998). *"Just tell me what to do" barriers to assessment-in-learning in higher education*. Paper presented to Scottish Educational Research Association Annual Conference, University of Dundee.
6. Güner, N. (2012). The effect of preventive classroom management training program on approval and disapproval behaviors of teachers. *Online Submission, 5*(1), 153–166.

7. Gunter, P. L., Venn, M. L., Patrick, J., Miller, K. A., & Kelly, L. (2003). Efficacy of using momentary time samples to determine on-task behavior of students with emotional/behavioral disorders. *Education and Treatment of Children, 26*(4), 400–412.
8. Harrop, A., & Daniels, M. (1986). Methods of time sampling: A reappraisal of momentary time sampling and partial interval recording. *Journal of Applied Behavior Analysis, 19*(1), 73–77.
9. Harrop, A., & McCann, J. (1983). Behaviour modification and reading attainment in the comprehensive school. *Educational Research, 25*, 191–195.
10. Harrop, A., & McCann, J. (1984). Modifying "creative writing" in the classroom. *British Journal of Educational Psychology, 54*, 62–72.
11. Harrop, A., & Swinson, J. (2000). Natural rates of approval and disapproval in British infant, junior and secondary classrooms. *British Journal of Educational Psychology, 70*(4), 473–483.
12. Jolly, M., & McNamara, E. (1992). *Pupil behaviour schedule towards better behaviour*, Towards Better Behaviour, Ainsdale.
13. Merrett, F., & Wheldall K. (1990). *Positive teaching in the primary school*. London: SAGE.
14. Nafpaktitis, M., Mayer, G. R., & Butterworth, T. (1985). Natural rates of teacher approval and disapproval and their relation to student behavior in intermediate school classrooms. *Journal of Educational Psychology, 77*(3), 362.
15. Schwieso, J., & Hastings, N. (1987). Teachers' use of approval. In N. Hastings & J. Schwieso (Eds.), *New directions in educational psychology, Vol. 2: Behaviour and motivation in the classroom* (pp. 115–136). London: Falmer.
16. Sulla, F., Perini, S., & Rollo, D. (2013). Analisi funzionale dei comportamenti di approvazione e disapprovazione nell'interazione insegnante-allievo: una rassegna. *Psicologia dell'Educazione, 7*(2), 193–216.
17. Swinson, J., & Harrop, A. (2001). The differential effects of teacher approval and disapproval in junior and infant classrooms. *Educational Psychology in Practice, 17*(2), 157–167.
18. Taras, M. (2003). To feedback or not to feedback in student self-assessment. *Assessment and Evaluation in Higher Education, 28*(5), 549–565.
19. Weinstein, L., Laverghetta, A., Alexander, R., & Stewart, M. (2009). Teacher greetings increase college students' test scores. *College Student Journal, 43*(2), 452–453.

MIX
Papier aus verantwortungsvollen Quellen
Paper from responsible sources
FSC® C105338

If you have any concerns about our products,
you can contact us on
ProductSafety@springernature.com

In case Publisher is established outside the EU,
the EU authorized representative is:
Springer Nature Customer Service Center GmbH
Europaplatz 3, 69115 Heidelberg, Germany

Printed by Libri Plureos GmbH
in Hamburg, Germany